ACTION AT A DISTANCE IN PHYSICS AND COSMOLOGY

A SERIES OF BOOKS IN ASTRONOMY AND ASTROPHYSICS

Editors: Geoffrey Burbidge and Margaret Burbidge

ACTION AT A DISTANCE IN PHYSICS AND COSMOLOGY

F. Hoyle
CALIFORNIA INSTITUTE OF TECHNOLOGY

J. V. Narlikar
TATA INSTITUTE OF FUNDAMENTAL RESEARCH

W. H. FREEMAN AND COMPANY
San Francisco

Library of Congress Cataloging in Publication Data

Hoyle, Fred.
Action at a distance in physics and cosmology.

Bibliography: p.
1. Quantum electrodynamics. 2. Gravitation.
3. Cosmology. I. Narlikar, Jayant Vishnu, 1938–
joint author. II. Title.
QC680.H69 530.1′2 74-4158
ISBN 0-7167-0346-7

Printed in the United States of America

1 2 3 4 5 6 7 8 9

CONTENTS

PREFACE

By the term "action at a distance," we mean a relativistically invariant particle interaction, not the instantaneous action at a distance of Newton. It is customary in physics to use such interactions in some problems, for example, scattering, but the usual point of view regards interparticle action as only a part of a wider interaction of fields and particles. Fields are taken to possess further degrees of freedom that are not contained within the interparticle action. Such wider degrees of freedom are called on in other kinds of problems, to explain the decay of excited atomic states, for example.

Our purpose in this book is to argue that all the results of physics, even those which are usually thought to arise from the independent degrees of freedom of fields, can be obtained from interparticle action—for decay problems and scattering problems alike. In developing this point of view, we have concentrated throughout on electrodynamics and gravitation, because these parts of physics, being more complete than particle physics is at present, would seem to provide the most stringent test of our thesis. At the end, however, we consider how an apparently critical difference between long-range and short-range interactions can be resolved—an issue relevant to extending the theory to strong and weak interactions.

The essential difference between field theory and what might be termed direct-interaction theory is that one can consider local systems in isolation from the rest of the universe in the former, but not in the latter. The independent degrees of freedom of fields in field theory are logically equivalent to the interaction of local systems with the universe in the direct-interaction theory. Since the equivalence is found to operate in certain cosmological models but not in others, the equivalence is by no means just a mathematical transformation of one theory into the other. We learn important aspects of the cosmological structure of the universe from it. "Cosmology" appears in the title of the book for this reason.

Although we have tried to present the argument in a self-contained way, not drawing extensively on outside results, the style presupposes for the most part that the reader is already familiar with "the usual theory," by which we mean field theory. The book is based on a series of lectures given at the NATO summer school, Erice, Sicily, 1972. We have incorporated the lecture format into the structure of the book because we found it permitted us to divide the treatment into fairly individual packets. Some authors apparently think it a virtue to write a book like a single long mathematical proof, where one cannot read a part without reading the whole. We ourselves find such expositions discouraging. In our experience, the most useful books are those in which it is possible to jump fairly freely from one place to another. Ideally, a book should be multidimensional in its construction. We hope the lecture format has helped somewhat toward this goal. Particularly, we have had in mind the reader who may wish to jump over the details of quantum electrodynamics, and perhaps over the formal aspects of Chapter 5. Such a reader, especially if interested more in gravitation and inertia than in electrodynamics, may wish to proceed directly from Chapter 3 to Chapter 6. This is why Chapter 6 begins in a more subdued vein than one might expect from the style of Chapters 4 and 5.

The authors wish to thank Rose Stratton and Evaline Gibbs for their help in preparing the manuscript. One of us (J. V. N.) is happy to acknowledge the award of a Jawaharlal Nehru fellowship.

F. H.

J. V. N. *1973*

ACTION AT A DISTANCE IN PHYSICS AND COSMOLOGY

1

INTRODUCTION

In this book we shall be concerned with seeing how the structure of the universe may be relevant to physics. The current and conventional point of view is that, although the universe probably sets important boundary conditions for the operation of physical laws, the laws themselves are independent of large-scale structure and could in principle be determined by entirely local experiments. This book takes the opposite point of view, that the physical laws, as we usually state them, already involve the universe as a whole.

In many problems it is possible to "decouple" the effect of the universe, in the sense that the influence of the universe remains effectively constant throughout the limited spacetime volumes that these problems concern. We shall, for example, regard particle masses as arising in the manner of Figure 1.1, which shows the paths of two particles, a and b. A biscalar propagator, satisfying a certain explicit wave equation, $\widetilde{G}(A, B)$, connects every pair of points A, B, on these paths. The mass of particle a, *at the point* A, is determined by a suitably defined summation of $\widetilde{G}(A, B)$ both with respect to variable B on the path of b and with respect to all particles b in the universe. It happens in many problems that the resulting mass, which can be denoted by $m_a(A)$, is almost completely determined by particles b that are very distant from a, and that $m_a(A)$ is effectively independent of the

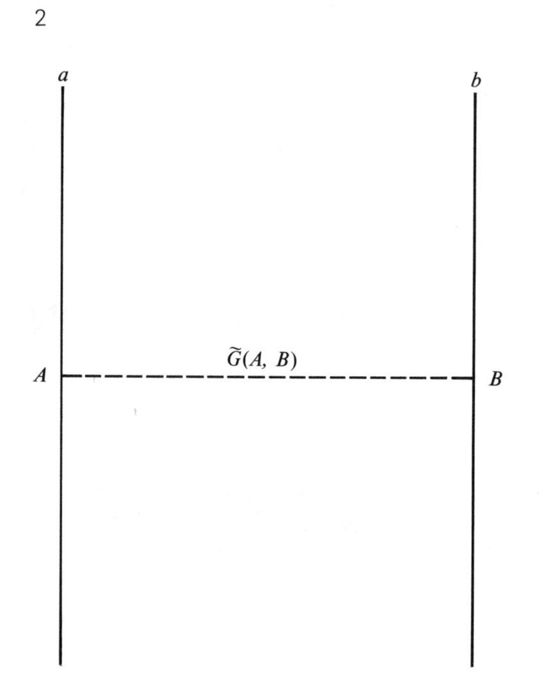

FIGURE 1.1
The mass of particle a at the point A arises from interactions with other particles b, the interaction being carried by the biscalar propagator $\tilde{G}(A, B)$, which is integrated over all points B on the world line of particle b.

position of A within localized spacetime volumes. This independence permits us to think of $m_a(A)$ as a constant—say, m—in these problems. In such cases, we can consider particle a to obey the usual localized rules, whether classical or quantum, as in the Dirac equation

$$(\nabla\!\!\!\!/ + im)\psi = 0; \qquad \nabla\!\!\!\!/ \equiv \gamma^i \frac{\partial}{\partial x^i},$$

where m is thought of as "belonging" to the particle. Such problems gain greatly in simplicity because the influence of the universe need not appear explicitly, provided we regard m as being empirically determined. This gain is at the expense of being able to determine m theoretically, however. Physics, working locally, seems incapable of explaining particle-mass values.

If the reader will grant for the moment that such a point of view is correct, it will be clear that problems in which the universe exercises a constant environmental influence are likely to be more tractable than problems in which the environmental influence is not constant. Problems where the influence of the universe can be replaced by certain empirical choices, as with particle masses, are exactly the problems in which the most spectacular advances are likely to be made. Since it is the sensible practice of the physicist to concentrate heavily on those problems where progress can be made, a situation is only too likely to arise in which all solved problems are of the decoupled kind. This situation in turn is only too likely to promote the point of view mentioned at the outset. It is easy to see that, by selecting for solvable problems, one could be led to the erroneous position of supposing that the physical laws were purely local—although many empirical choices would be needed to make such laws "work," just as many empirical choices are indeed made in modern physics.

The decoupling of local systems from the universe is not always as conceptually simple as it is for particle masses. The manner in which decoupling must be made in order to obtain correct results may be well-known, as it is in thermodynamics, but a sense of mystery remains. Such parts of physics are regarded as containing something uncomfortably odd.

In classical physics, and in quantum problems involving particles but not antiparticles, we are concerned with calculating how a state P, given at time t, changes into a state Q at time $t + \Delta t$, where Δt is a specified interval. The states P, Q, depend on the particle positions and momenta, and possibly on certain field quantities. (In quantum theory there are of course amplitudes for these variables, whereas in classical theory the variables take explicit values.) Provided the laws governing the change from P to Q are time-symmetric, one can obtain from Q a state $R(Q)$ with the property that a system with state R at time t goes to P at time $t + \Delta t$. For example, in classical mechanics $R(Q)$ is obtained from Q by reversing the signs of the momenta of the particle. Hence the class of all "final" states at time $t + \Delta t$ contains all "initial" states at time t. No time-sense can be determined in such a system. There are no irreversible processes.

Why then in the world of experience do we find the irreversibility of thermodynamics? A poor answer, but one that has often been given, argues that, because we cannot (for lack of technique) work through all the details that are involved in going from P to Q for many-particle systems of great complexity, some new mysterious factor has been overlooked. The argument is that, were it possible to make such explicit calculations, a miraculous principle, not apparent for simpler systems, would emerge.

A second answer, one that has a closer relation to the world of experience, turns on restricting the class of initial states. Practical thermodynamics considers the state P to be that of a local system, not of the whole universe. The variables that define P therefore depend in part on the local system itself and in part on the manner in which the system is related to the outside world, and this dependence inevitably restricts the variables to the values which are actually possible in our particular universe. This restriction means that we do not have access to the whole class of all possible initial states but only to a subclass, say, $\mathcal{K}$. This subclass of states at time t defines a subclass, say, $\mathcal{L}$, of states at time $t + \Delta t$. and these two subclasses need not be the same. Thus a P belonging to $\mathcal{K}$ going to a Q belonging to $\mathcal{L}$ may define an $R(Q)$ that does not happen to be realizable in our universe, i.e. which does not belong to $\mathcal{K}$. We now have a time-sense, but one that owes its origin to the restrictions imposed by the universe.

A third answer is to deny the time-symmetry of the basic laws. The slight difference in the decay rates of the K^0, $\bar{K}^0$, mesons represents a small departure from time-symmetry. If this asymmetry is interpreted as basic to the physical laws, one might seek to relate all other asymmetries, such as that of thermodynamics, to the behavior of the K^0, $\bar{K}^0$, mesons. We ourselves regard this as a possible, but not a promising, approach to the problem. We think it better to seek to explain the K^0, $\bar{K}^0$, difference in terms of a particle-antiparticle asymmetry of the universe.

So far as we are aware, no other suggestion for explaining thermodynamic irreversibility has been proposed. Of these three suggestions, only the second seems at first sight to provide firm ground. Yet there is a most important aspect of the third suggestion, connected not with mesons but with electromagnetism, which must also be considered. Although the statement of the laws of electromagnetism is fully time-symmetric, all practical applications of the laws are asymmetric. The laws are made to yield asymmetric results by means of an *ad hoc* restriction to only time-retarded solutions of Maxwell's equations. This curious circumstance presents us with two alternatives. First, we can argue that the choice of time-retarded solutions is a basic principle of physics which we do not seek to explain. We then face two possibilities, which depend on whether we consider the empirically observed asymmetry of the universe to be derived from, or independent of, this new principle. The second possibility would require us to attribute thermodynamic irreversibility to a combination of two independent asymmetries, one belonging to the universe, the other to the time-retarded principle. The first possibility would require us to attribute all asymmetries, even those of the universe, to time-retardation.

The second alternative is quite different in principle, and amounts to inverting this first possibility: we can attribute both thermodynamic irreversibility and time-retardation to the structure of the universe—and this is the point of view we follow in this book. This alternative has the attraction that the need for imposing time-retardation as a basic principle is obviated. Surprisingly, we shall find that time-*symmetric* solutions of the electrodynamic equations can yield all the observed effects, provided local problems are properly related to the universe and provided the universe has an appropriate large-scale structure. This last requirement shows that, although the theory we will develop yields all the usual results, of both classical and quantum electrodynamics, it is not equivalent to the usual theory. The usual theory does not tell us anything about the large-scale structure of the universe.

Much of what has just been said has been known, at least for classical electrodynamics, from work of Wheeler and Feynman done more than twenty years ago. The classical theory already shows how the unwanted advanced solutions of Maxwell's equations are canceled by what Wheeler and Feynman called the "response of the universe." Our particular task will be to show how those quantum phenomena which are usually taken to arise from the zero-point fluctuations of the quantized electromagnetic field can also be explained in a fully time-symmetric theory in terms of the response of the universe. Much of the first half of the book is concerned with this task. Readers who are more interested in sections dealing with the masses of particles and with general relativity may prefer to pass rather lightly over the quantum electrodynamics. However, we consider the discussion of the latter to be crucial, since we would not like to elaborate the concepts of this book unless they were first shown to be capable of recovering the best-developed part of modern physics.

So far, we have discussed two important aspects of the world, thermodynamic irreversibility and the need to choose retarded solutions of Maxwell's equations, both of which fit uneasily into a purely local theory. Are there other practical issues which also fit uneasily, or not at all, into a local theory?

Many scientists have remarked on a curious apparent coincidence involving the physical constants and the Hubble constant H determined from the redshift-distance relation

$$z = \frac{\Delta\lambda}{\lambda} = Hd$$

taken for galaxies whose distances d are not so large that the regions of spacetime in question are markedly non-Euclidean. Here $\Delta\lambda$ is the wavelength shift for a spectrum line of laboratory wavelength λ. The coincidence in question appears most clearly when the velocity of light c and Planck's constant $\hbar$ are set equal to unity, so that redundant units are suppressed. With $c = \hbar = 1$, spatial lengths and time intervals have the same unit, and particle masses become inverse lengths. The length unit is indeed best determined from the mass of some particle, from the $\pi^{\pm}$ mesons, for example, as, say, m_{π}^{-1}. All physical quantities then have dimensionalities which can be expressed in terms of a combination of m_{π} with certain dimensionless numbers. The latter arise either from particle mass ratios or from particle interactions which serve to define four coupling constants, α_1, α_2, α_3, α_4. For strong interactions, $\alpha_1 \simeq 15$; $\alpha_2 = e^2 = \frac{1}{137}$ is the fine-structure constant equal to the square of the elementary charge; $\alpha_3 \simeq 10^{-5}$ for weak interactions; and

$$\alpha_4 = Gm_\pi{}^2 \simeq 10^{-40}$$

for the gravitational interaction, G being the Newtonian constant. Physical processes involve integral powers, not fractional powers, of these coupling constants; so the surprise of finding a physical constant as small as α_4 cannot be avoided simply by replacing α_4 by a small fractional power of α_4. The "coincidence" can now be stated. It is

$$H/m_\pi \simeq \alpha_4,$$

meaning that the ratio of the local length scale $m_\pi{}^{-1}$ to the cosmological length scale H^{-1} is also of order α_4. Because the usual local formulation of physics throws no light on this "coincidence," many modern cosmologists have eschewed an attempt to understand it. For our part, however, we believe the apparent coincidence of these numbers to be another indication that physics should not be formulated in a purely local way.

As another example of an issue that fits uneasily into a local theory, we mention here the concept of an "origin" for the universe. The meaning of this concept will be a major topic in the second half of the book.

2

CLASSICAL THEORIES OF ELECTROMAGNETISM AND GRAVITATION

§2.1. HISTORICAL DEVELOPMENTS (LECTURE 1)

Theoretical physics, as we know it today, really began with Newton's laws of motion and gravitation. The remarkable work of Copernicus, Kepler, Galileo, and others before them had provided a wealth of observational material on phenomena in the solar system and on the Earth. These seemed to show that nature follows some definite set of rules and patterns. Could these, in turn, be explained by simpler and fewer assumptions? An affirmative answer was given by Newton's laws. Phenomena that had seemed diverse, like the pendulum, the falling apple, and the motion of planets, were shown to follow the same set of laws.

A wrong law often explains all existing phenomena, but breaks down completely as new phenomena are uncovered. Newton's laws, however, were able to surmount this hurdle successfully for two centuries, and consequently came to dominate the course of theoretical physics until the end of the nineteenth century. An example of this dominance is found in the origin of the electromagnetic theory, and it is interesting to follow the development of this theory alongside that of gravitation.

The law of gravitation appears as an action-at-a-distance statement, in which particles with masses m_1, m_2, attract each other by a force

$$F = G\frac{m_1 m_2}{r^2}, \tag{1}$$

where r is their distance apart and G a constant. Coulomb's law of attraction/repulsion between two electric charges with magnitudes e_1, e_2, or magnetic poles μ_1, μ_2, distance r apart, follows the same pattern:

$$F \propto \pm\frac{e_1 e_2}{r^2}, \qquad F \propto \pm\frac{\mu_1 \mu_2}{r^2}, \tag{2}$$

where $+$ denotes attraction and $-$ denotes repulsion.

Repulsion as well as attraction is possible because charges and poles can exist in two species. In this respect (2) differs from (1): this difference is crucial, and led to very different developments of gravitation and electromagnetism. Since all masses attract gravitationally, bigger masses tend to dominate smaller masses. Therefore one is necessarily led to study the interactions of large bodies, like planets, satellites, or stars. To detect influences of gravitation on a laboratory scale is extremely difficult and requires very sensitive equipment. The situation is reversed for electromagnetism, since large objects like planets or stars are electromagnetically neutral, and since it is easy to produce isolated charges and magnetic dipoles in the laboratory. Thus (2) was studied very extensively in the laboratory.

Although initially there was a good agreement between (2) and the laboratory experiments, more sophisticated experiments began to show that (2) does not represent the whole truth. In particular, phenomena associated with sufficiently rapidly moving charges are not explained by (2). It is true that (2) can be patched up to make it agree with experiment by adding further terms to the righthand side, but the resulting expression looks very inelegant. Moreover, the similarity with (1) which initially led to (2) is then lost.

The solution of the electromagnetic problem came in an altogether different way, with the introduction of the so-called electromagnetic field. This entity is supposed to surround every electric charge and convey an influence from one charge to another. For example, when charge e_1 moves, it disturbs the field in its neighborhood. This disturbance propagates outward with *finite* velocity and eventually reaches charge e_2. The disturbance is then communicated to e_2, which begins to move. The propagation of such disturbances in the field was described by waves. The action-at-a-distance picture came therefore to be replaced by one of particles and fields. The mathematical structure for this was provided by Maxwell.

The finite velocity of propagation in the electromagnetic field was identified experimentally with the velocity of light. Because of their phenomenal success in describing electromagnetism, Maxwell's equations inspired a revolution in physics. In 1905 Einstein proposed the special theory of relativity, which recognized c, the velocity of light, as having a special significance in being the maximum velocity attainable by any physical entity. This recognition led to a revision of the ideas of spacetime and motion that had been held sacrosanct by the physicist since Newton.

Special relativity explained why the law given by (2) failed to describe all electromagnetism. The interaction, according to (2), is instantaneous; i.e., (2) assumes that the interaction propagates from one particle to another with infinite speed. This assumption is refuted by the rule that c is the maximum velocity possible in our universe.

This conclusion naturally led people to wonder whether (1) was also wrong! Once we accept special relativity, we must object to (1) on the same grounds that apply to (2). Indeed, if we were able to experiment with rapidly moving objects in the laboratory in order to test (1), the discrepancy would long ago have been noticed. The massive astronomical objects available for study do not move fast enough to reveal any such discrepancy. So objections to (1) must be based on theoretical grounds.

It is possible to modify (1) in several different ways to make it consistent with relativity. Einstein himself proposed a radical revision of the law of gravitation in 1915, which is known as the general theory of relativity. We therefore have the following interrelations between theories of gravitation and of electromagnetism:

Newton's laws of gravitation and motion
↓
Coulomb's laws of electromagnetism
↓
Maxwell's equations (electromagnetism)
↓
Special relativity (motion)
↓
General relativity (gravitation)

In this chapter we shall examine Maxwell's equations and general relativity to see how successful they have been in this century.

§2.2. MAXWELL-LORENTZ ELECTRODYNAMICS: THEORETICAL FRAMEWORK

Maxwell's equations can be written most elegantly in a four-dimensional format. We define a vector field A_i $(i = 1, 2, 3, 4)$ to be the 4-potential, and derive the electromagnetic field F_{ik} by skew-differentiation:

$$F_{ik} = \frac{\partial A_k}{\partial x^i} - \frac{\partial A_i}{\partial x^k}. \tag{3}$$

Here x^i, $i = 1, 2, 3, 4$, are the spacetime coordinates, with x^1, x^2, x^3, representing Cartesian space coordinates, and $x^4 = ct$ the time coordinate. We shall assume $c = 1$ in all formal discussion, and use c explicitly only where we are concerned with experiment or observation. Special relativity defines an invariant line element

$$ds^2 = \eta_{ik}\, dx^i\, dx^k, \tag{4}$$

where $\eta_{ik} = \text{diagonal}\,(-1, -1, -1, +1)$; η_{ik} is used for raising or lowering indices.

Maxwell's equations consist of two sets. The first one is the identity implicit in (3):

$$F_{ik;l} + F_{kl;i} + F_{li;k} \equiv 0, \tag{5}$$

where a semicolon denotes differentiation with respect to coordinates. The second set is given by

$$F^{ik}{}_{;k} = -4\pi J^i, \tag{6}$$

where J^i is the 4-current vector. The equations (5) and (6) permit a gauge transformation

$$\widetilde{A}_i = A_i + \phi_{;i}, \tag{7}$$

where ϕ is a scalar. We can choose ϕ such that

$$\widetilde{A}^i{}_{;i} = 0. \tag{8}$$

Henceforth we shall assume that such a transformation has been made and use (8) without the tilde; i.e.,

$$A^i{}_{;i} = 0. \tag{9}$$

It is then easy to see that

$$\Box A^i \equiv A^{i;k}{}_{;k} = 4\pi J^i, \tag{10}$$

which shows that the electromagnetic disturbances in vacuo propagate as waves with velocity $c\,(= 1)$.

To make connection with the older experimental data, we relate the above entities to the pre-relativity quantities—the electric and magnetic 3-vectors $\boldsymbol{E}$ and $\boldsymbol{H}$, the current density $\boldsymbol{j}$, and the charge density ρ—by means of the following identifications:

$$\begin{aligned}
(A^1, A^2, A^3) &\equiv \boldsymbol{A} \text{ (the vector potential)},\\
A^4 &\equiv \phi \text{ (the scalar potential)},\\
(J^1, J^2, J^3) &\equiv \boldsymbol{j}, \qquad J^4 \equiv \rho,\\
(F_{23}, F_{31}, F_{12}) &\equiv -\boldsymbol{H},\\
(F_{41}, F_{42}, F_{43}) &\equiv \boldsymbol{E}.
\end{aligned} \tag{11}$$

We then get the familiar Maxwell equations in a medium of unit dielectric constant and permeability:

$$\begin{aligned}
\nabla \cdot \boldsymbol{E} &= 4\pi\rho, & \nabla \cdot \boldsymbol{H} &= 0,\\
\nabla \times \boldsymbol{H} &= 4\pi\boldsymbol{j} + \frac{\partial \boldsymbol{E}}{\partial t}, & \nabla \times \boldsymbol{E} &= -\frac{\partial \boldsymbol{H}}{\partial t}.
\end{aligned} \tag{12}$$

The gauge condition becomes

$$\nabla \cdot \boldsymbol{A} + \frac{\partial \phi}{\partial t} = 0. \tag{13}$$

$\boldsymbol{A}$ and ϕ satisfy the wave equations

$$\frac{\partial^2 \boldsymbol{A}}{\partial t^2} - \nabla^2 \boldsymbol{A} = 4\pi \boldsymbol{j}, \qquad \frac{\partial^2 \phi}{\partial t^2} - \nabla^2 \phi = 4\pi\rho. \tag{14}$$

So far we have equations describing how fields are related to the distribution of charges and currents. To complete the picture we need equations of motion of charges in electromagnetic fields. These are supplied by the so-called Lorentz force equations. In the 4-dimensional format, these are

$$m_0 \frac{du^i}{ds} = eF^i{}_k u^k, \tag{15}$$

where m_0 is the rest mass and e the charge of a particle moving in the field F_{ik} with 4-velocity u^i. If $\boldsymbol{v}$ is the corresponding Newtonian 3-velocity and m the mass, we have

$$m = \frac{m_0}{\sqrt{1 - v^2}}, \tag{16}$$

and in three-dimensional notation, (15) becomes

$$\frac{d}{dt}(m\boldsymbol{v}) = e[\boldsymbol{E} + \boldsymbol{v} \times \boldsymbol{H}]. \tag{17}$$

The expression on the righthand side is known as the Lorentz force.

The Maxwell-Lorentz picture, described above in its simplest form, has worked remarkably well in classical physics. More details can be introduced to describe properties of various materials, but the basic picture stands. The theory had to be modified when the quantum nature of the universe became apparent; yet quantum electrodynamics based on the Maxwellian picture was still able to explain many of the detailed properties of matter.

Nevertheless, the theory has certain unsatisfactory features which tend to suggest that it is far from perfect. The objections range from purely formal ones to computational ones at high energies. We shall review some of these in the following sections.

Retarded and Advanced Solutions

The electromagnetic problem, according to the Maxwell-Lorentz theory, consists of two parts. One, as shown by (10) or (14), is to compute the electromagnetic fields due to a given charge-current distribution. The other part is to compute the motion of charged particles in given electromagnetic fields according to equations (15) or (17). We shall consider the first part in this section and the second part in the following section.

To solve (10) we make use of Green's functions. Consider a solution $G(X, X_1)$ of the scalar wave equation

$$\Box G(X, X_1) = \delta_4(X - X_1), \tag{18}$$

where $\delta_4(X - X_1)$ is the four-dimensional Dirac delta function and the $\Box$ operator is with respect to coordinates of point X. The details of such solutions are discussed extensively in the literature. For the present we shall assume that a solution exists.

Consider now a four-dimensional domain V surrounding a point X_1, and bounded by a 3-surface Σ. Using Green's theorem,

$$\int_\Sigma [A_i G_{;k} - A_{i;k} G] n^k \, d\Sigma = \int_V [A_i G_{;k} - A_{i;k} G]^{;k} \, dV,$$

where the differentiations are with respect to X, which is a variable point. Thus we have

$$\begin{aligned}\int [A_i(X)\, G(X, X_1)_{;k} - A_i(X)_{;k}\, G(X, X_1)] n^k \, d\Sigma \\ &= \int [A_i \,\Box G - \Box A_i \, G] \, dV \\ &= A_i(X_1) - \int 4\pi J_i(X)\, G(X, X_1) \, dV.\end{aligned}$$

Hence

$$A_i(X_1) = 4\pi \int G(X, X_1)\, J_i(X) \, dV + \int [A_i(X)\, G(X, X_1)_{;k} - A_i(X)_{;k}\, G(X, X_1)] n^k \, d\Sigma. \quad (19)$$

We are therefore able to express the value of A_i at an interior point in terms of the source J_i in V and the values of A_i and its normal derivative on the boundary Σ of V. This result is known as Kirchhoff's theorem.

So far we have not said anything about the form of G. We now choose an explicit solution of (18) given by

$$G = \frac{1}{4\pi|\boldsymbol{x} - \boldsymbol{x}_1|} \delta[t_1 - t - |\boldsymbol{x} - \boldsymbol{x}_1|], \quad (20)$$

where $(\boldsymbol{x}, t)$ and $(\boldsymbol{x}_1, t_1)$ are the coordinates of X and X_1, respectively. It can be verified that (20) indeed satisfies (18).

From (19) we now see that the contribution to A_i at $(\boldsymbol{x}_1, t_1)$ comes from only those points which satisfy

$$t_1 - t - |\boldsymbol{x} - \boldsymbol{x}_1| = 0, \quad (21)$$

i.e., from points on the past light cone of X_1. The values of J_i or A_i in the future do not appear to make any contribution at all. In particular, if we choose V as the slab of spacetime

$$t_2 \leq t \leq t_3 \qquad (t_2 < t_1 < t_3), \quad (22)$$

the contribution to A_i comes only from the past light cone of X_1 bounded by $t = t_2$ (see Figure 2.1). If further we let $t_2 \to -\infty$, and assume that the contribution of the second integral in (19) tends to zero, we are left with just the volume integral. Then (19) can be written in the form

$$A(x_1, t_1) = \int \frac{j(x, t_1 - |x - x_1|)}{|x - x_1|} d^3x,$$

$$\phi(x_1, t_1) = \int \frac{\rho(x, t_1 - |x - x_1|)}{|x - x_1|} d^3x. \tag{23}$$

The solution obtained in (23) is known as the *retarded* solution, for the obvious reason that the potentials at (x_1, t_1) are determined by the charge-current distributions at *earlier* times. If we had retained the second integral in (19) over $t = t_2$, but *not* over $t = t_3$, we would have additional contributions from the values of A and ϕ specified at an earlier time t_2. This addition of the second integral also does not violate the retarded nature of the solution.

The retarded solution appeals to our intuition because it conforms to the usual idea of causality. If we imagine the electromagnetic disturbances as generated by charge movements, then we expect these disturbances to arise

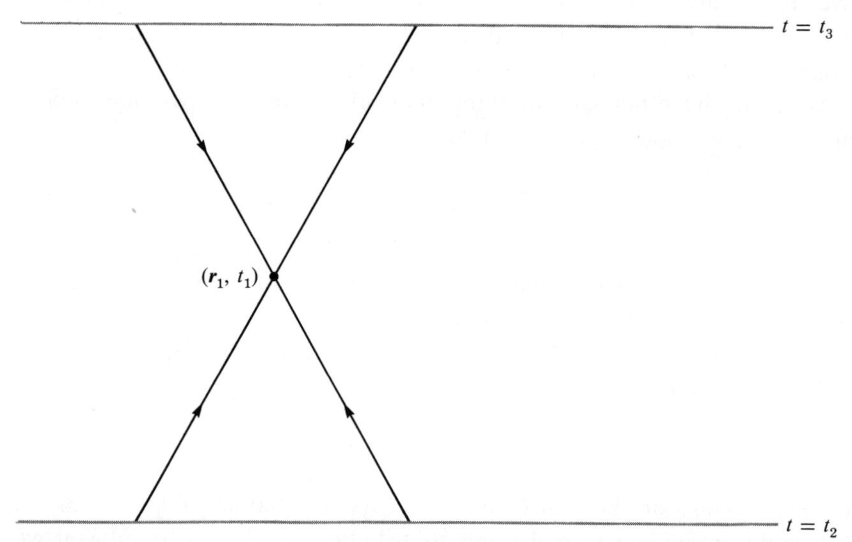

FIGURE 2.1
Advanced and retarded Green's functions. Retarded contributions to $(\mathbf{r}_1, t_1)$ come from the past hypersurface $t = t_2$. Advanced contributions come from the future hypersurface $t = t_3$.

subsequent to the motion of the charges. This is expressed by the retarded solutions (23). It would seem, at first sight, that we have achieved the remarkable result of *explaining* causality through Maxwell's equations.

Such a conclusion is erroneous, however, as can be seen from the fact that we obtained (23) with the use of (20) for G, and this is not a unique procedure. We could have chosen other forms of G, with different results. For example, if we had taken

$$G = \frac{1}{4\pi|\boldsymbol{x} - \boldsymbol{x}_1|}\,\delta[t - t_1 + |\boldsymbol{x} - \boldsymbol{x}_1|], \tag{24}$$

which is also a solution of (18), we would have arrived at

$$\boldsymbol{A}(\boldsymbol{x}_1, t_1) = \int \frac{\boldsymbol{j}(\boldsymbol{x}, t_1 + |\boldsymbol{x} - \boldsymbol{x}_1|)}{|\boldsymbol{x} - \boldsymbol{x}_1|}\, d^3\boldsymbol{x},$$

$$\phi(\boldsymbol{x}_1, t_1) = \int \frac{\rho(\boldsymbol{x}, t_1 + |\boldsymbol{x} - \boldsymbol{x}_1|)}{|\boldsymbol{x} - \boldsymbol{x}_1|}\, d^3\boldsymbol{x}, \tag{25}$$

instead of (23), together with surface integrals, which might vanish. In (25) the disturbances in the electromagnetic fields appear to arise *prior* to the motion of the charges! This solution is called the *advanced* solution.

Clearly causality is violated by (25). But since (25) also satisfies Maxwell's equations, we cannot deduce causality from Maxwell's equations. Indeed, these equations are invariant under time reversal,

$$t \rightarrow -t, \tag{26}$$

and so for any retarded solution there must also exist an advanced solution, and vice versa.

We may ask which of the two solutions, (23) or (25), is the correct one. To answer this question, we go back to (19), which retains the surface integral. Provided we thus retain the surface integral, either way of describing A_i is correct. When we apply (19) to the slab $t_2 \leq t \leq t_3$ and let $t_2 \rightarrow -\infty$ or $t_3 \rightarrow \infty$, the same conclusion holds, provided we retain the surface integrals. However, when we make the assumption that one or both of the surface integrals over $t = t_2$, $t = t_3$, vanishes, we are introducing a new element which is *not* contained in the Maxwell equations. It is this element, which is a boundary condition, that makes the solution take the form (23) or (25). In order to be consistent with causality, we choose the boundary condition leading to (23) rather than (25). For Maxwell's equations themselves, the choice is arbitrary. The equations supply us with both advanced and retarded solutions (and, because of their linearity, with any linear combination of them). Of these, we select the one in agreement with experience.

With so many solutions theoretically possible, why does nature always select the retarded one? That this question cannot be answered within the framework of Maxwell's theory must be regarded as one of its intrinsic weaknesses.

The Paradox of Self-Action

We now turn to the influence of fields on charged particles, as expressed by (17). Although (17) poses no problems so long as the fields $\boldsymbol{E}$, $\boldsymbol{H}$, are due to other sources, a curious situation arises when we include in $\boldsymbol{E}$ and $\boldsymbol{H}$ the fields produced by the charge itself. Theoretically, these fields must also act on the charge. If, following the arguments of the last section, we take these fields as retarded ones, they carry energy from the charge. By conservation of energy, we expect the charge to lose energy, and the "self" force to damp the motion of the charge.

Using Maxwell's equations, we can evaluate the retarded fields produced by the moving charge. To obtain the self force, we need this field taken at the charge itself. Now, for a point charge the field is infinite, but the self force is finite, because it arises from the averaging of the self field in all directions. In the averaging process, many infinite terms cancel, and we are left with a force term which is finite. The limiting process is an involved one; see Dirac (1938a) for details. We shall here try to reproduce some of Dirac's ideas in a simple, nonrelativistic way.

Consider an electric charge e moving in a straight line in an external field E in the x-direction. The equation of motion is

$$m\ddot{x} = eE + \frac{2e^2}{3}\dddot{x}, \tag{27}$$

where the last term is the self-force term obtained after the limiting operation described above. It is easy to see that if the charge is oscillating, the self-force damps the motion. For the rate of work done by the self-force,

$$\frac{2}{3}e^2\dot{x}\,\dddot{x} = \frac{2}{3}e^2\left[\frac{d}{dt}(\dot{x}\,\ddot{x}) - \ddot{x}^2\right]. \tag{28}$$

The time-average of the first term on the righthand side of (28) vanishes in an oscillatory motion, whereas the second term is negative definite. Thus the charge loses its kinetic energy at an average rate of

$$\frac{2}{3} e^2 \overline{\ddot{x}^2}. \tag{29}$$

If we calculate the electromagnetic waves generated by the charge, we find that they carry energy across a sphere at infinity at the same average rate (29). This verifies the conservation of energy mentioned above. At first sight, then, the self-force appears to act correctly. However, a paradox begins to appear when we apply (27) to the case $E = 0$. We then have

$$\ddot{x} = \lambda \dddot{x}, \qquad \lambda = \frac{2e^2}{3m}. \tag{30}$$

This has the trivial solution $\dot{x}$ = constant. It also has an exponential solution,

$$\dot{x} = (\text{constant}) \exp(t/\lambda). \tag{31}$$

Although $\dot{x}$ = constant causes no problem, the exponential solution leads to unbounded velocities as $t \to \infty$. Nor can the solution (31) be avoided. For example, if in practice a charge passes from a region of non-zero electromagnetic field to a region of zero field, it will not in general be true that all higher derivatives $\ddot{x}, \dddot{x}, \ldots$, are zero on entry to the zero field. Hence the solution (31) is generated.

To avoid unbounded solutions, Dirac suggested the following procedure. Consider a simple example where the charge is unaffected by any external force except for an impulse at $t = 0$. The equation of motion for this case can be written in the form

$$\ddot{x} = P\,\delta(t) + \lambda \dddot{x}, \tag{32}$$

where P is the strength of the impulse. Now, a conventional solution of (32) would be:

$$\begin{aligned} \dot{x} &= 0, && \text{for } t < 0; \\ \dot{x} &= P(1 - e^{t/\lambda}), && \text{for } t > 0. \end{aligned} \tag{33}$$

As mentioned before, this solution is unbounded. Instead of (33), Dirac proposed the following solution:

$$\begin{aligned} \dot{x} &= P \exp(t/\lambda), && \text{for } t < 0; \\ \dot{x} &= P, && \text{for } t > 0. \end{aligned} \tag{34}$$

Notice that, because of the presence of the $\dddot{x}$ term in (32), $\ddot{x}$ rather than $\dot{x}$ is discontinuous at $t = 0$ in both solutions. The second solution is bounded for all times. However, this boundedness has been achieved at a cost. In (34), although the impulse is applied at $t = 0$, the charge begins to move earlier, and builds up a velocity and acceleration that the impulse just cancels at $t = 0$. This anticipatory movement of the charge violates causality. Dirac argued that this breakdown of causality may be due to the internal structure of the electron. However, to a physicist accustomed to the usual ideas of causality, the solution (34) will be unpalatable.

Yet the conventional solution (33) also has conceptual as well as mathematical difficulties. One may ask, "Where does the charge get its kinetic energy if its velocity becomes larger and larger according to (33)?" The answer lies in the self-field of the particle. The energy of this field is infinite for a point charge, and the particle is able to draw on its source indefinitely. This is no more palatable than the acausality of (34).

It seems, therefore, that infinite self-interaction cannot be avoided in classical Maxwellian electrodynamics. With the introduction of quantization, it was hoped to circumvent these difficulties. Although it does turn out that quantum theory reduces the order of this infinity, it cannot altogether remove it. Furthermore, quantum theory introduces another infinity of non-classical origin, the vacuum polarization. Elaborate renormalization procedures are required to extract meaningful finite results from apparently divergent calculations.

These considerations suggest that all is not well with Maxwellian electrodynamics, in either its classical or its quantum form. On the other hand, its impressive experimental successes suggest that the theory does represent some considerable measure of truth and should not be lightly abandoned. In Chapter 3 we shall investigate another form of electrodynamics, which resembles Maxwell's in its experimental details, but which is better able to meet the type of criticism we have raised.

§2.3. GENERAL RELATIVITY: THEORETICAL FRAMEWORK (LECTURE 2)

Einstein approached the subject of gravitation in a remarkable way. He argued that gravitation is intimately connected with the geometric properties of spacetime. To follow this argument, it is helpful to compare and contrast the properties of gravitation with those of the electromagnetic field. From the broad similarity of Newton's and Coulomb's laws, we might expect gravitation to be a field in the Maxwellian sense. However, the electromagnetic field is capable of being switched on or off. The absence of any observed "anti-gravity" in nature suggests that the same is not true of gravitation. Thus the gravitational field in a given region of spacetime is more permanent than the electromagnetic field.

Einstein expressed this result by relating gravitation to the spacetime structure itself, that is, by postulating that the presence of gravitation means the spacetime geometry is non-Euclidean. In the special theory of relativity, the invariant line element

$$ds^2 = \eta_{ik}\, dx^i\, dx^k$$

plays a fundamental part. General relativity requires (4) to be modified to take account of gravitation. In general we write

$$ds^2 = g_{ik}\, dx^i\, dx^k, \tag{35}$$

where x^i ($i = 1, 2, 3, 4$) are not necessarily Cartesian coordinates, although (35) retains a critical property of (4): the signature of the quadratic form. Although g_{ik} can now be real functions of x^i, the form (35), when diagonalized at any point, must have as a signature either $(- - - +)$ or $(+ + + -)$. We shall use $(- - - +)$ throughout.

In special relativity, (4) is invariant under Lorentz transformations. In general relativity, (35) is invariant under general coordinate transformations $x^i \rightarrow x'^i(x^k)$, where x'^i are twice-differentiable functions of x^k.

The effect of coordinate transformations is to cause vectors and tensors to undergo transformations also. A contravariant vector A^i transforms according to

$$A'^i = \frac{\partial x'^i}{\partial x^l} A^l, \tag{36}$$

and a covariant vector B_k as

$$B'_k = \frac{\partial x^l}{\partial x'^k} B_l. \tag{37}$$

Tensors are defined by appropriately extending these laws.

The geometric properties of spacetime are those of a Riemannian manifold. Thus we define parallel displacement of vectors A^i, B_k, through an infinitesimal coordinate change δx^l by

$$\delta A^i = -\Gamma^i{}_{kl} A^k\, \delta x^l, \qquad \delta B_k = \Gamma^i{}_{kl}\, B_i\, \delta x^l, \tag{38}$$

where

$$\Gamma^i{}_{kl} = \frac{1}{2} g^{im} \left(\frac{\partial g_{mk}}{\partial x^l} + \frac{\partial g_{ml}}{\partial x^k} - \frac{\partial g_{kl}}{\partial x^m} \right), \tag{39}$$

the matrix with elements g^{im} being reciprocal to the matrix $\|g_{im}\|$.

The covariant derivatives of A^i and B_k are given by

$$A^i{}_{;l} = \frac{\partial A^i}{\partial x^l} + \Gamma^i{}_{kl} A^k, \qquad B_{k;l} = \frac{\partial B_k}{\partial x^l} - \Gamma^i{}_{kl} B_i. \tag{40}$$

The "non-Euclidean" nature of the geometry is apparent when we displace A^i or B_k parallel to itself along a closed curve. It fails to return to its starting value! Quantitatively this failure is expressed by the noncommutativity of repeated covariant differentiations:

$$A^i{}_{;kl} - A^i{}_{;lk} = R^i{}_{mlk} A^m, \tag{41}$$

and a similar relation for B_k. The tensor $R^i{}_{mlk}$ is the Riemann-Christoffel tensor. It is given by

$$R^i{}_{mlk} = \frac{\partial \Gamma^i{}_{mk}}{\partial x^l} - \frac{\partial \Gamma^i{}_{ml}}{\partial x^k} + \Gamma^i{}_{nl} \Gamma^n{}_{km} - \Gamma^i{}_{nk} \Gamma^n{}_{ml}. \tag{42}$$

From $R^i{}_{mlk}$ we can construct two lower-rank tensors:

$$R_{mk} = R^i{}_{mki}, \qquad R = g^{mk} R_{mk}. \tag{43}$$

R_{mk} is called the Ricci tensor and R the scalar curvature. The combination

$$G_{mk} = R_{mk} - \frac{1}{2} g_{mk} R \tag{44}$$

has zero divergence:

$$G^{mk}{}_{;k} = 0. \tag{45}$$

G_{mk} is the Einstein tensor.

These relations characterize the geometric properties of spacetime. Physical ideas are introduced through Einstein's equations of gravitation:

$$G_{ik} = -\kappa T_{ik}, \tag{46}$$

where T_{ik} is the energy-momentum tensor of matter and of all physical interactions, and κ is a coupling constant chosen to be

$$\kappa = 8\pi G, \tag{47}$$

where G is the Newtonian constant. The choice (47) is dictated by the requirement that the gravitational equations reduce to the Newtonian form in the limit of weak fields, i.e., when g_{ik} differs from η_{ik} by quantities of the first order of smallness. The details of this analysis are found in standard textbooks on relativity, and we shall not repeat them here. We point out in passing that, because of (45), $T^{ik}{}_{;k} = 0$; i.e., the law of conservation of energy follows from Einstein's equations.

Gravitational Collapse: Black Holes

Gravitation possesses an exceedingly curious property, which can be illustrated by comparing the gravitational attraction between two masses with the attraction between two particles connected by a stretched elastic string. In the latter case, the particles approach each other until the string attains its natural length; the attraction ceases thereafter. In the former case, however, the particles approach each other with ever-increasing attraction—until they bump into each other. The difference in the two cases is reflected in the potential energy stored in the two systems. In the elastic-string case, the potential energy is given by

$$V = \lambda(r - a)^2, \tag{48}$$

where λ is a constant, r is the separation of the particles and a is the natural length of the string. In the gravitational case, the potential energy, obtained from Newton's law (1), is

$$V = -\frac{Gm_1m_2}{r}. \tag{49}$$

The expression (48) is positive definite and is characteristic of most systems in the world. But (49), on the other hand, is negative, and goes to $-\infty$ as $r \to 0$. In both cases the system seeks to attain the position of minimum potential energy. Whereas with (48) it is able to do so in a smooth, steady way, with (49) the system undergoes catastrophic collapse. Of the various known interactions in physics, gravitation appears to be unique in exhibiting this property. The fact that (49) is negative is often described by the statement that the energy stored in a gravitational field is negative.

This discussion is based on Newton's law of motion, and one wonders whether general relativity is different in this respect. As it turns out, general relativity leads to an even more catastrophic situation, which we shall discuss in some detail, since it has occupied the attention of physicists for the last decade.

We discuss first a simple problem using Newtonian laws. Consider a uniform spherical object of mass M, density ρ_0 and radius R_0 which begins contraction under its own gravitation. We assume no internal pressures. Then a particle on the surface of the object, of radius R at time t, has as equation of motion

$$\ddot{R} = -\frac{GM}{R^2}, \tag{50}$$

with $R = R_0$, $\dot{R} = 0$ at time $t = 0$. It is easy to see that (50) gives

$$\dot{R}^2 = 2GM\left(\frac{1}{R} - \frac{1}{R_0}\right); \tag{51}$$

so the object shrinks to a point in a time τ,

$$\tau = \int_0^{R_0} \sqrt{\frac{RR_0}{2GM(R_0 - R)}}\, dR = \frac{\pi}{2\sqrt{\alpha}}, \tag{52}$$

where

$$\alpha = \frac{2GM}{R_0^{\,3}} = \frac{8\pi}{3} G\rho_0. \tag{53}$$

Thus a uniform object with the mean density of the Sun, ~ 2 gm cm^{-3}, would shrink to a point in about 25 minutes if all its internal pressure were withdrawn.

This Newtonian calculation is very closely simulated by general relativity. It is found that the interior of the object can be described by the line element

$$ds^2 = dt^2 - S^2(t)\left[\frac{dr^2}{1 - \alpha r^2} + r^2(\sin^2\theta\, d\phi^2 + d\theta^2)\right], \qquad r \leq R_0, \tag{54}$$

where r, θ, ϕ, are spherical polar coordinates that are taken constant for a freely falling particle, and t is the proper time of the particle. The exterior solution, $r > R_0$, is given by

$$ds^2 = dt^2 - S^2\left(\frac{tR_0^{3/2}}{r^{3/2}}\right)\left[\frac{K^2\,dr^2}{1 - (\alpha R_0^3/r)} + r^2(d\theta^2 + \sin^2\theta\,d\phi^2)\right],$$

$$K = \frac{1}{S}\frac{\partial}{\partial r}(rS), \tag{55}$$

in the same "comoving coordinates." The function $S(t)$ in the interior represents the fraction by which the object has contracted at any given time t. If we define $R(t) = R_0S(t)$, then $R(t)$ satisfies the same equation (51).

Thus, according to general relativity, the object collapses into $R = 0$ in the same time as that given by (52). However, the situation here is more drastic, for not only does the body fall into a point, but spacetime itself becomes singular at $t = \pi/(2\sqrt{\alpha})$!

At first one might imagine the following way out of this situation. Consider an observer exterior to the object, and not falling into it. Such an observer can work in terms of the exterior Schwarzschild line element,

$$ds^2 = dT^2\left(1 - \frac{2GM}{\rho}\right) - \frac{dR^2}{1 - \dfrac{2GM}{\rho}} - \rho^2(d\theta^2 + \sin^2\theta\,d\phi^2), \tag{56}$$

where

$$\rho = rS, \qquad T = T(r, t). \tag{57}$$

If such an observer uses light signals to maintain contact with an observer who is falling freely with the object, the following situation develops. So long as the freely falling observer satisfies the condition

$$rS(t) > 2GM, \tag{58}$$

any signals transmitted by him will eventually be received by the external observer. Contact is broken, however, when (58) is violated, as it eventually must be. As this stage is approached, the signals take longer and longer in T-time to pass from one observer to the other, and the violation of (58) is never discovered by the external observer, no matter how long he lives. When (58) is violated, the object is supposed to have become a "black hole"; i.e., it is so highly collapsed that its strong gravitational field prevents any escape of light.

Thus no spacetime singularity at $r = 0$ is ever visible to the external observer. All he can ever see is an object collapsing toward a black hole.

This apparent resolution of the singularity problem is not satisfactory, however, since it excludes the experience of the freely falling observer. A more satisfactory resolution would require spacetime to be nonsingular everywhere.

The collapse of the object can be halted in Newtonian gravitation by putting sufficiently strong pressures in the object. The same prescription does not work in general relativity, however. Any agency causing pressure has energy, and energy attracts, according to relativity. Thus the self-gravitation of the object is increased by the very agency that is meant to stop it. These general concepts were quantitatively expressed first by Penrose (1965), who showed that under certain general conditions no "reasonable" physical agency is capable of halting gravitational collapse into a singularity. A "reasonable" agency is one which has positive energy in the sense discussed earlier in this section.

Is the singularity of gravitational collapse an academic one? Recent astronomical discoveries of quasars and pulsars suggest the contrary. Astronomers are beginning to observe massive, highly condensed objects in which self-gravitation is likely to be the predominant force. The continued stable existence, and sometimes outward expansion, of these objects suggests that the ideas of general relativity might require modification for strong gravitational fields.

Cosmological Models: Singularity

The singularity described above also appears in the cosmological solutions of Einstein's equations. The simplest of the cosmological models are the homogeneous and isotropic ones described by the Robertson-Walker line element,

$$ds^2 = dt^2 - S^2(t)\left[\frac{dr^2}{1 - kr^2} + r^2(d\theta^2 + \sin^2\theta \, d\phi^2)\right]. \tag{59}$$

Here t is the cosmic time and r, θ, ϕ, the spherical polar coordinates of a typical galaxy—or a fundamental observer. The parameter k is 0 or ± 1. The spaces $t =$ constant are homogeneous and isotropic. They are flat if $k = 0$, have positive curvature and finite volume if $k = +1$, and negative curvature if $k = -1$. $S(t)$ is a scale factor. For a dust model, the energy-momentum tensor is given by

$$T^{ik} = \rho u^i u^k, \tag{60}$$

where $\rho = \rho(t)$, and $u^i = (0, 0, 0, 1)$ corresponds to the motion of a fundamental observer. The Einstein equations then lead to

$$\rho S^3 = \text{constant} = A \text{ (say)}, \tag{61}$$

and

$$3\,\frac{\dot{S}^2 + k}{S^2} = 8\pi G\rho = \frac{8\pi GA}{S^3}. \tag{62}$$

The similarity between (54) and (59), and between (51) and (62), is obvious, and indeed the solution for a finite object (51) is identical with that for the universe if $k = 1$. The only difference is that for a finite object $R(t)$ was decreasing with t, whereas for an expanding universe $S(t)$ increases with t. The singularity at $S = 0$ is the same as that discussed earlier. The cases $k = 0, -1$, also have the singular state $S = 0$. This state is described in cosmological studies as the "big bang."

Work by Hawking and Penrose (1969) has shown that the singularity $S = 0$ is inevitable in all cosmological models under certain quite general conditions. As with a finite object, one of these conditions is the positive definiteness of the energy of all physical systems contained in T_{ik}. This condition, expressed by

$$T_{ik}\, v^i v^k > 0, \tag{63}$$

where v^i is any timelike vector, is satisfied by all interactions in physics known so far. We are therefore led to a catastrophic situation for the universe similar to that for a finite collapsing object.

The cases discussed here and in the previous section suggest that something may well be wrong with a straightforward extrapolation of our knowledge from weak gravitational fields to strong ones. This possibility is recognized by many theoretical physicists, although they differ in their approaches to solving the problem.

§2.4. ACTION PRINCIPLE FOR ELECTROMAGNETISM AND GRAVITATION (LECTURE 3)

The Maxwell and Einstein equations discussed so far took their present form only after many experiments and much trial and error. It turns out, however, that they can be derived in a simple and elegant way from the principle of stationary action. Since we shall be concerned with this principle in the new approach to these subjects, it is worth repeating the derivation of the Maxwell and Einstein equations from an action principle.

The action is given by

$$S = \frac{1}{16\pi G}\int R\sqrt{-g}\,d^4x - \sum_a \int m_a\,da$$

$$-\frac{1}{16\pi}\int F_{ik}F^{ik}\sqrt{-g}\,d^4x - \sum_a \int e_a A_i\,da^i. \quad (64)$$

The various quantities appearing in (64) have all been defined before. In particular, e_a, m_a, refer to the charge and mass of a typical particle denoted by a. The variable parameters in (64) are A_i and g_{ik} and the world lines of the particles. We obtain the required physical equations by setting $\delta S = 0$ for small arbitrary variations of these parameters.

It is easy to verify that variation of A_i leads to Maxwell's equations in curved space:

$$\Box A_i \equiv -F_i{}^k{}_{;k} = 4\pi J_i, \quad (65)$$

where

$$J^i(X) = \sum_a e_a \int \frac{\delta_4(X, A)}{\sqrt{-g(X)}}\frac{da^i}{da}\,da \quad (66)$$

is the current vector.[1]

Variation with respect to geometry g_{ik} is more cumbersome, but eventually leads to

$$R^{ik} - \frac{1}{2}g^{ik}R = -8\pi G(T^{ik}{}_{(m)} + T^{ik}{}_{(em)}), \quad (67)$$

where $T^{ik}{}_{(m)}$ is the matter tensor,

$$T^{ik}{}_{(m)}(X) = \sum_a \int m_a \frac{\delta_4(X, A)}{\sqrt{-g(X)}}\frac{da^i}{da}\frac{da^k}{da}\,da, \quad (68)$$

and $T^{ik}{}_{(em)}$ is the electromagnetic tensor,

[1]The gauge being chosen so that $A^k{}_{;k} = 0$.

$$T^{ik}{}_{(em)} = \frac{1}{4\pi}\left(\frac{1}{4}g^{ik}F_{mn}F^{mn} - F^{il}F^{km}g_{lm}\right). \tag{69}$$

Finally, variation of the world line of a leads to the equation of motion of a:

$$\frac{d^2a^i}{da^2} + \Gamma^i{}_{kl}\frac{da^k}{da}\frac{da^l}{da} = \frac{e_a}{m_a}F^i{}_k\frac{da^k}{da}. \tag{70}$$

For an uncharged particle we get the geodesic equation.

The three variations are actually interdependent, as may be seen by taking the divergence of (67) in the neighborhood of a particle a. By taking appropriate limits, and using (65), (66), we arrive at (70), because the variation of a world line can be described in terms of variation of g_{ik} and of A_i. This point is considered in some detail in Chapter 8.

Conformal Invariance

We end this chapter with a brief discussion of conformal transformations, which will play an important part in later work.

Consider a spacetime manifold M in which the metric is given by g_{ik}. If we now change g_{ik} to a new metric $g^*{}_{ik}$, given by, say,

$$g^*{}_{ik} = \Omega^2 g_{ik} = e^{2\zeta}g_{ik}, \tag{71}$$

we have a new spacetime M^*. When Ω is a finite nonzero function of the coordinates, (71) is called a conformal transformation. It is easy to see that any element of length ds in M transforms to ds^* in M^*, where

$$ds^* = \Omega\, ds = e^{\zeta}\, ds. \tag{72}$$

Thus the ratio of two such lengths in different directions, but at the same point, remains unchanged under a conformal transformation.

A quantity is said to be conformally invariant if it does not change under (71). One example is the electromagnetic action. It is easy to see that if we put

$$A^*{}_i = A_i, \tag{73}$$

then the electromagnetic part of (64) is equal to a similar expression in terms of $A^*{}_i$ and $F^*{}_{ik}$ with respect to M^*. Variation of $A^*{}_i$ in this invariant part of the action leads to Maxwell's equations with respect[2] to F^*_{ik} and to M^*. These equations are said to be invariant under conformal transformations. We shall consider later whether conformal invariance is a property that should be satisfied by other parts of physics.

The space M is said to be conformally flat if we can find a coordinate system in which

$$g^*_{ik} = \eta_{ik},$$
$$g_{ik} = \Omega^{-2}\,\eta_{ik}. \tag{74}$$

Later we shall show that the Robertson-Walker spaces given by (59) are conformally flat.

Einstein's equations are not conformally invariant. It can be verified that under the conformal transformation (71) R_{imlk}, R_{ik}, and R transform as

$$R^*_{imlk} = e^{2\zeta}[R_{imlk} + g_{ik}\zeta_{ml} + g_{ml}\zeta_{ik} - g_{il}\zeta_{mk} - g_{mk}\zeta_{il} + (g_{ik}g_{ml} - g_{il}g_{mk})\,\Delta_1\zeta], \tag{75}$$

$$R^*_{ik} = R_{ik} + 2\zeta_{ik} + g_{ij}(\Box\zeta + 2\,\Delta_1\zeta), \tag{76}$$

$$R^* = e^{-2\zeta}(R + 6\Box\zeta + 6\,\Delta_1\zeta) = \Omega^{-2}(R + 6\Omega^{-1}\Box\Omega), \tag{77}$$

where

$$\zeta_{ij} = \zeta_{;ij} - \zeta_{;i}\zeta_{;j}, \qquad \Delta_1\zeta = \zeta_{;i}\zeta^{;i}. \tag{78}$$

From (75), (76), (77), we can construct another tensor which is invariant under (71). This is the Weyl conformal curvature tensor:

$$C^h{}_{ijk} = R^h{}_{ijk} + \frac{1}{2}[g^h{}_jR_{ik} - g^h{}_kR_{ij} + g_{ik}R^h{}_j - g_{ij}R^h{}_k + \frac{1}{6}R(g^h{}_kg_{ij} - g^h{}_jg_{ik})]. \tag{79}$$

[2] But not with respect to the wave equation for A^*_i. If one wishes to obtain $\Box A^*_i = 4\pi J^*_i$, it is necessary to add a gauge transformation in order that $A^{*k}{}_{;k} = 0$, i.e., to define A^*_i by $A^*_i = A_i + \phi_{;i}$, where the scalar function ϕ satisfies an equation of the form given in (26) of Chapter 5.

3

THE ABSORBER THEORY OF WHEELER AND FEYNMAN

§3.1. THE FOKKER ACTION (LECTURE 4)

In 1845, in a letter to Weber, Gauss wrote about action at a distance in electrodynamics: "I would doubtless have published my researches long ago were it not that, at the time I gave them up, I had failed to find what I regarded as the keystone, *Nil actum reputans si quid superesset agendum:* namely, the derivation of the additional forces—to be added to the interaction of electrical charges at rest, when they are both in motion—from an action which is propagated not instantaneously but in time, as is the case with light."

Gauss's problem was solved in the early decades of the present century by several physicists, notably Schwarzschild, Tetrode, and Fokker. Essentially the problem was to make action at a distance relativistically invariant. We give here the formulation devised by Fokker, in which the action is

$$S = -\sum_{a} \int m_a \, da - \sum_{a<b} \sum \int \int e_a e_b \, \delta(s^2{}_{AB}) \, da^i \, db^k \, \eta_{ik}, \tag{1}$$

where

$$s^2{}_{AB} = (a^i - b^i)(a^k - b^k) \, \eta_{ik}, \tag{2}$$

a^i, b^i, being the 4-coordinates of typical points A and B on the world lines of particles a, b, respectively. We are here regarding the particles as being ordered numerically, to give meaning to $a < b$. The notation implies that in the double sum the pair a, b, is counted only once.

Thus the charges interact if and only if A and B can be connected by a light ray. In the language of special relativity, "action at a distance" is an unfortunate description. We really have action at zero 4-dimensional distance, $s^2{}_{AB} = 0$.

It is instructive to compare S with the action for the Maxwellian theory given in Chapter 2. The second term of (1) includes all the electromagnetic effects of the Fokker theory, and should be compared with both the pure field term and the interaction term of Maxwell's theory. To show that (1) in fact describes the same electrodynamic properties of matter as Maxwell's fields do, we define the following quantities:

$$A_i^{(b)}(X) = e_a \int \delta(s^2{}_{XB})\, \eta_{ik}\, db^k, \tag{3}$$

$$F_{ik}^{(b)} = A_{k;i}^{(b)} - A_{i;k}^{(b)}, \tag{4}$$

$$J_i^{(b)}(X) = e_b \int \delta_4(X, B)\, \eta_{ik}\, db^k. \tag{5}$$

$A_i^{(b)}$ and $F_{ik}^{(b)}$ may be called the potential and field arising from particle b, whereas $J_i^{(b)}$ is the 4-current density represented by the motion of the charge b. Because of Dirac's identity,

$$\Box_X\, \delta(s^2{}_{AX}) \equiv \eta^{ik}\, \delta(s^2)_{;ik} \equiv 4\pi\, \delta_4(A, X), \tag{6}$$

we get

$$\Box A_i^{(b)}(X) \equiv 4\pi\, J_i^{(b)}(X). \tag{7}$$

Also, if the charge e is neither created nor destroyed, i.e., if its world line stretches from $t = -\infty$ to $t = +\infty$, we get

$$A^{(b)i}{}_{;i} \equiv 0. \tag{8}$$

It follows from (7), (8), and (4) that

$$F^{(b)ik}{}_{;k} \equiv -4\pi J^{(b)i}. \tag{9}$$

The similarity between (7), (8), (9), and the Maxwell equations now becomes more apparent. It should, however, be remembered that, unlike the Maxwell equations, (7), (8), and (9) are identities because of the definitions (3), (4), (5). The fields ${F^{(b)}}_{ik}$ or the potentials ${A^{(b)}}_i$ are defined in terms of the particle world line; they have no degrees of freedom of their own. To distinguish them from ordinary fields, we shall call them direct-particle fields whenever there is likely to be any confusion.

The variation of the world line of a leads to the equation of motion of charge a:

$$m_a \frac{d^2 a^i}{da^2} = e_a \sum_{b \neq a} {F^i}_k{}^{(b)} \frac{da^k}{da}. \tag{10}$$

This again is similar to the Lorentz force equation, but with the difference that the field of the particle a itself is specifically excluded from the righthand side. There is no "self action."

This completes the formal description of the theory of direct interparticle action according to Fokker. Although superficially there is similarity with Maxwell's theory, this theory is different in one crucial practical respect. As the action functional (1) indicates, there is complete time-symmetry. This is shown more explicitly if we work out the potential (3). It has the form

$$A_i^{(b)} = \frac{1}{2} A_{i(\mathrm{ret})}{}^{(b)} + \frac{1}{2} A_{i(\mathrm{adv})}{}^{(b)}, \tag{11}$$

where $A_{i(\mathrm{ret})}{}^{(b)}$ is the retarded solution of Maxwell's equations for the same charge, and $A_{i(\mathrm{adv})}{}^{(b)}$ is the advanced solution. The light cone from a general point X intersects the world line of b in two points. The motion of b at the point on the past part of the cone defines the retarded solution, whereas the motion at the point on the future part defines the advanced solution (see Figure 3.1).

As discussed in Chapter 2, our experience is consistent with retarded solutions, and therefore in disagreement with the above formulation. For this reason, although the theory so far developed surmounted the difficulty of the lack of Lorentz invariance of the Coulomb theory, it still seemed a long way from being a successful rival to Maxwell's theory. It was not until 1945 that the next important development in the theory took place.

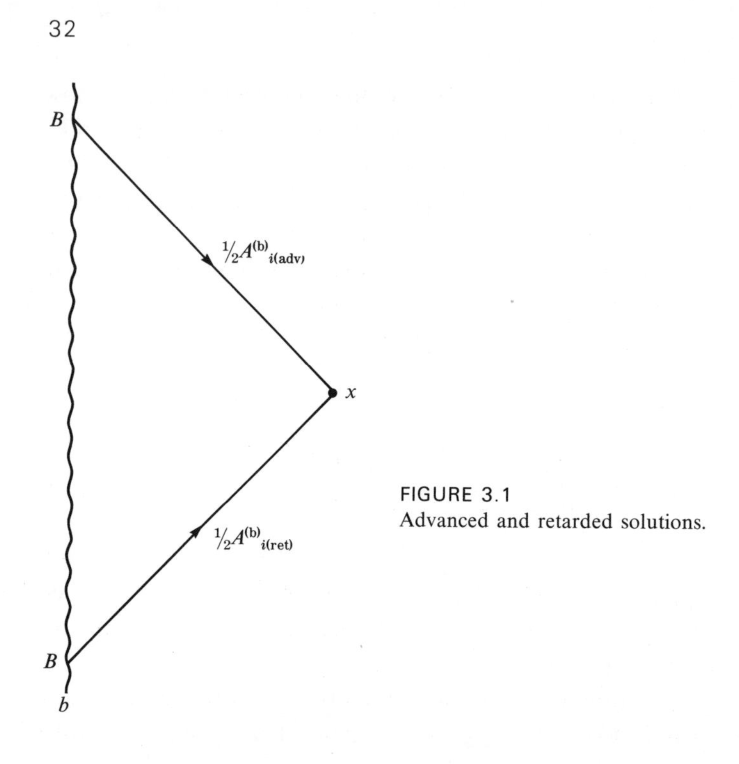

FIGURE 3.1
Advanced and retarded solutions.

§3.2. THE RESPONSE OF THE UNIVERSE (LECTURE 5)

The presence of advanced solutions side by side with the retarded ones can lead to paradoxical situations. Let us describe one as a typical example.

Suppose two observers, *A* and *B*, situated one light-hour apart, can communicate with each other via advanced as well as retarded light signals. They agree to set up a signaling system in which *A* sends a signal to *B* at 4 P.M. if and only if he does *not* hear from *B* at 4 P.M., and *B* sends a signal to *A* at 5 P.M. if and only if he does hear from *A* at that time. Will this arrangement work? In a world of retarded signals only, it can work. But a little reasoning tells us that it is unworkable in the present situation. For, if *A* does send a signal at 4 P.M., it reaches *B* at 3 P.M. and 5 P.M. So *B* sends a signal at 5 P.M. which reaches *A* at 4 P.M. and 6 P.M. But if *A* does get a signal at 4 P.M., he should not send one out at 4 P.M. The other alternative is also impossible.

That a theory with advanced and retarded signals can be made to work in the real world was first realized by Wheeler and Feynman. They pointed out that in the real world one has to take into account a "response of the universe." How the universe enters the argument in a crucial way can be seen as follows. Suppose we move an electric charge a at time $t = 0$. Its retarded effect is felt by another charge b at distance r at time $t = r$. The advanced reaction of b on a reaches the charge a at time $t = r - r = 0$! Thus the reaction is instantaneous, no matter how large r may be. To find out the electromagnetic effects of moving charge a, we must therefore take into account the reactions of all other charges b in the universe. The combined reaction of all these charges is termed the response of the universe. Wheeler and Feynman gave the following simple calculation of the response in a static Euclidean universe with a uniform distribution of charges of different species.

Suppose we move the charge a, and describe its general motion by $\boldsymbol{a}(t)$. The velocity and acceleration of a are $\dot{\boldsymbol{a}}(t)$ and $\ddot{\boldsymbol{a}}(t)$, respectively. Since the calculation is linear throughout, we can proceed by Fourier analysis. Let a typical Fourier component of the acceleration be given by

$$\boldsymbol{u}_0\, e^{-i\omega t}. \tag{12}$$

To simplify the picture, Wheeler and Feynman imagined the charge a located at the center of a spherical cavity of radius R. The cavity is supposed to be empty except for charge a. The medium outside the cavity is uniform, with N charges e_k per unit volume.

In vacuum, the full retarded electric field of a at large distance r (see Figure 3.2) would be

$$u_0 \frac{e}{r} \sin\theta \exp[i\omega(r - t)] \tag{13}$$

in the direction of increasing θ, where e is the charge of a, and θ is the angle between the radius vector and the acceleration. Modifying (13) to take

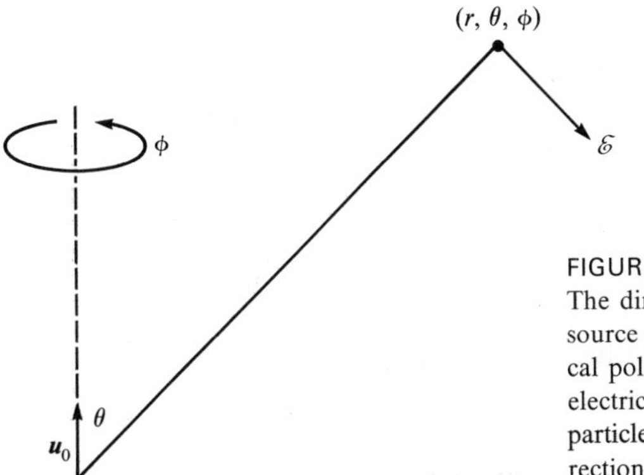

FIGURE 3.2
The direction $\boldsymbol{u}_0$ of the acceleration of the source particle gives the axis of the spherical polar coordinates r, θ, ϕ. The retarded electric field in vacuum due to the source particle has magnitude $\mathcal{E}$ and is in the direction of increasing θ.

account of the medium, first we have a factor

$$2(1 + n - ik)^{-1} \tag{14}$$

arising from reflection at $r = R$, where $n - ik$ is the refractive index of the medium. We also have an additional phase factor

$$\exp[i\omega(n - ik - 1)(r - R)] \tag{15}$$

due to transmission in the medium. The electric field in the present problem is therefore

$$\mathscr{E} = \frac{2eu_0 \sin\theta}{r(1 + n - ik)} \exp\{i\omega[r - t + (n - ik - 1)(r - R)]\} \tag{16}$$

in the direction of increasing θ. This field acts on a charge b to produce an acceleration

$$\frac{e_k}{m} p(\omega)\mathscr{E} \tag{17}$$

also in the direction of increasing θ. Here m is the mass, e_k is the charge of b, and $p(\omega)$ is a frequency-dependent factor which includes damping effects. The function $p(\omega)$ is related to the refractive index by

$$(n - ik)^2 = 1 - \frac{4\pi N e_k{}^2}{m\omega^2} p(\omega). \tag{18}$$

The motion of charge b produces a half-advanced field at a, which constitutes the reaction of b on a. The component of this field parallel to $\boldsymbol{u}_0$ has the magnitude

$$\frac{e_k}{m} p(\omega) \left(\frac{e_k}{2r} \sin\theta\right) \exp(-i\omega r)\mathscr{E}. \tag{19}$$

The net response $\mathscr{R}$ of the universe is obtained by summing (19) over all charges $b \neq a$,

$$\mathscr{R} = \int_{r=R}^{\infty} \int_{\theta=0}^{\pi} \int_{\phi=0}^{2\pi} \frac{e_k{}^2}{2rm} p(\omega) \sin\theta\, e^{-i\omega r}\mathscr{E} \cdot Nr^2 \sin\theta\, dr\, d\theta\, d\phi. \tag{20}$$

Straightforward integration, using (16) and (18), leads to

$$\mathscr{R} = -\frac{2}{3} i\omega\, eu_0\, e^{-i\omega t}. \tag{21}$$

A similar calculation shows that the component of the response normal to $\boldsymbol{u}_0$ averages to zero, so that (21) can be written in the form

$$\mathcal{R} = -\frac{2e\,i\omega}{3}\boldsymbol{u}_0\,e^{-i\omega t}. \tag{22}$$

We now go back to the Fourier sum of (12). It is clear from (22) that the response for the sum will be given by

$$\frac{2e}{3}\dddot{\boldsymbol{a}}, \tag{23}$$

and the charge experiences a force of reaction given by

$$\frac{2e^2}{3}\dddot{\boldsymbol{a}}. \tag{24}$$

If we had calculated the response field not at $r = 0$, but in the neighborhood of $r = 0$ in the cavity, we would have found that it amounts to

$$\frac{1}{2}F^{(a)}{}_{\text{ret}} - \frac{1}{2}F^{(a)}{}_{\text{adv}}, \tag{25}$$

where $F^{(a)}{}_{\text{ret}}$ is a typical retarded-field component of a. This calculation, which is slightly more involved than above, was also given by Wheeler and Feynman. They argued that (25) modifies the time-symmetric field of a, namely

$$\frac{1}{2}F^{(a)}{}_{\text{ret}} + \frac{1}{2}F^{(a)}{}_{\text{adv}}, \tag{26}$$

to produce the full retarded field observed in practice.

The above calculation is a self-consistent one. We start with the assumption that the net electromagnetic disturbance (13) from the particle is the full retarded one. We then show how this disturbs other charges and we calculate their response. The response must come out to be that given by (25) for the calculation to be self-consistent.

It is worth emphasizing how the two parts of this calculation differ in the use of the refractive index. In the first part, we anticipate that the net field in the neighborhood of each particle will be the retarded one. As this travels outward, it disturbs other charges whose field, as a result of this disturbance, must be added to the original field. This many-particle process is described by the use of the refractive index. In the second part of the calculation, we calculate the advanced reaction produced by each charge b on a as determined by the action terms

$$-e_a e_b \int\int \delta(s^2{}_{AB})\, da_i\, db^i, \tag{27}$$

and this two-particle interaction has no refractive index involved.

The Absorber Theory of Radiation

The remarkable fact which emerges from this calculation is that the final result (22) does not involve details of the rest of the universe, e.g., parameters like N or $n - ik$. These quantities appear in the intermediate steps, but are neatly canceled out in the end. This cancelation is not a coincidence, but involves an important property of the universe: the property of perfect absorption, which means that when the charge a is excited, all electromagnetic disturbances arising from it—directly or through the reaction of other charges—should tend to zero sufficiently rapidly at great distances from a. Mathematically, we say that

$$\sum_b \frac{1}{2}[F^{(b)}{}_{\text{ret}} + F^{(b)}{}_{\text{adv}}] \sim o\left(\frac{1}{r}\right) \qquad \text{as } r \to \infty. \tag{28}$$

In a vacuum a radiative field drops off as r^{-1}. What (28) indicates is a faster rate of attenuation than r^{-1}, to include genuine absorption. A universe which satisfies (28) is said to be a perfect absorber. Wheeler and Feynman demonstrated that charges influence one another by means of only retarded interactions in such a universe. Their proof is along the following lines.

The condition (28) refers to a combination of outgoing and incoming waves, and for it to hold at all times, we need to have the two types of waves tend to zero separately:

$$\sum_b \frac{1}{2} F^{(b)}{}_{\text{ret}} \sim o\left(\frac{1}{r}\right), \qquad \sum_b \frac{1}{2} F^{(b)}{}_{\text{adv}} \sim o\left(\frac{1}{r}\right), \qquad r \to \infty. \tag{29}$$

The conditions (29) also imply that

$$\sum_b \frac{1}{2}[F^{(b)}{}_{\text{ret}} - F^{(b)}{}_{\text{adv}}] \sim o\left(\frac{1}{r}\right) \qquad \text{as } r \to \infty. \tag{30}$$

However, the lefthand side of (30) is a field without sources and satisfies the homogeneous wave equation. The condition (30) at infinity therefore implies that it is identically zero everywhere:

$$\sum_b \frac{1}{2}[F^{(b)}{}_{\text{ret}} - F^{(b)}{}_{\text{adv}}] \equiv 0. \tag{31}$$

The field acting on charge a can thus be written in the form

$$\sum_{b\neq a}\left[\frac{1}{2}F^{(b)}{}_{\text{ret}} + \frac{1}{2}F^{(b)}{}_{\text{adv}}\right] = \sum_{b\neq a} F^{(b)}{}_{\text{ret}} + \frac{1}{2}[F^{(a)}{}_{\text{ret}} - F^{(a)}{}_{\text{adv}}]. \tag{32}$$

The first term on the righthand side represents the retarded effects of all charges $b \neq a$. The second term is identical to that obtained by direct calculation above. This completes the proof.

The second term on the righthand side of (32) is interesting in that it brings out the difference in approach to radiation of the Maxwell field theory and of the direct particle theory. The form of this term was first suggested by Dirac, who pointed out that its limiting value at the charge a is equal to the so-called radiative damping force. In field theory this term has no further significance, and Dirac's result remains somewhat mysterious. We have already noted that the radiative damping force is a self force in the Maxwell field theory. But in the above Wheeler-Feynman approach this term arises as a response of the universe. A charge experiences this force because of the complete absorption of its outgoing field. While absorbing a fraction of this disturbance, each absorber particle sends a reaction. The total reaction from the whole universe obtained by summing all absorber particles then has the above form.

This latter point of view has the advantage of preventing the runaway solutions discussed in Chapter 2, for if the charge gets on to such a solution, it will have an unbounded field which cannot be absorbed; so the formula leading to this solution will break down. The runaway solution is therefore not self-consistent within the framework of the theory.

The derivation of the general result (32) leaves one point somewhat unclear. At what stage have we introduced time-asymmetry into the argument? Clearly the condition (31) is also time-symmetric, but (32) is apparently not. Suppose we rewrite (32) in equivalent form,

$$\sum_{b\neq a}\left[\frac{1}{2}F^{(b)}{}_{\text{ret}} + \frac{1}{2}F^{(b)}{}_{\text{adv}}\right] = \sum_{b\neq a} F^{(b)}{}_{\text{adv}} + \frac{1}{2}[F^{(a)}{}_{\text{adv}} - F^{(a)}{}_{\text{ret}}], \tag{33}$$

with the "advanced" and "retarded" terms interchanged. There is nothing to prevent us from using (33) instead of (32). To do so within a calculation of the kind which led to (22) would be exceedingly awkward, however. In deriving (32) we assumed the absorber particle to be at rest *before* it was hit by the retarded source field. This is correct in the case of (32) because the first term on the righthand side is uncorrelated with the motion of a, but in (33) the first term is highly correlated. If we take these correlations into account, we discover that (33) is the same as (32).

But could we not make a similar self-consistent calculation in which the absorber particle came to rest *after* the advanced source field hit it? Within the Wheeler-Feynman theory such a time-reversed calculation can certainly be made. It is then the first term on the righthand side of (33) that is uncorrelated with the motion of a, and that on the righthand side of (32) is now highly correlated. This situation represents an inversion of the electromagnetic arrow of time. Thus there are self-consistent calculations leading to either sense of causality. Wheeler and Feynman sought to avoid this ambiguity by an appeal to thermodynamics. It was argued that for an absorber particle to come to rest *after* being hit by the source field would require highly "artificial" initial conditions—conditions with negligible statistical probability. Thus according to Wheeler and Feynman the time-asymmetry in electrodynamics is connected with the problem of initial conditions and thermodynamics.

§3.3. COSMOLOGICAL CONSIDERATIONS (LECTURE 6)

The original calculation of Wheeler and Feynman was for a static Euclidean universe. Such a universe is time-symmetric—unless one introduces asymmetric initial conditions, as mentioned above. In 1962 Hogarth criticized the Wheeler-Feynman approach on the grounds that it did not properly include the time-asymmetry of the large-scale structure of the universe. In an expanding universe, the future and the past have in general different physical properties. Over and above this, the advanced and retarded electromagnetic waves undergo changes during propagation. The former are blueshifted, the latter redshifted. Thus in an ever-expanding universe, the absorption of retarded waves is an interaction of low-frequency radiation with matter, but the absorption of advanced waves is an interaction at high energy. Thus the two cycles of argument, one based on retarded electromagnetic waves, the other on advanced waves, which take the same form in a static universe, become different in an expanding universe. Hogarth analyzed these problems in certain simple cosmological models, and was able to classify these models according to the nature of wave propagation. Hoyle and Narlikar (1963) and Roe (1969) later looked at the same problem from a different point of view. Their conclusions are essentially the same as those of Hogarth. We shall follow the later treatment here.

Before applying cosmology, we need to mention a formal problem concerning the definition of Fokker action. So far it has been defined in flat spacetime. In order to apply it to the Riemannian spacetime used in cosmology, we must write the action in a generally covariant form. Hogarth sought to avoid this step by using the conformal invariance of Maxwell's equations. Since Robertson-Walker spaces are conformally flat, he was able to use flat-space solutions for electromagnetic fields straight away in such spaces. More strictly, however, one should demonstrate (a) that the Fokker action can be written in a Riemannian space, and (b) that the action so written is conformally invariant. We shall look at these and other related formal problems in Chapter 5. For now, we can assume that the required results can be proved.

We begin by obtaining a general result analogous to that proved by Wheeler and Feynman in flat space. We show that in a cosmological model, for the retarded field to be consistent, it must be suitably attenuated at infinity—i.e., the future part of the universe must act as a perfect absorber. For the advanced field to be consistent, the past part of the universe must be a perfect absorber. We give the proof for retarded waves. That for advanced waves is similar.

Suppose we move charge a and consider the fields arising from such a motion. They are determined by

$$\frac{1}{2}[F^{(a)}{}_{\text{ret}} + F^{(a)}{}_{\text{adv}}] + \frac{1}{2}\sum_{b \neq a} [F^{(b)}{}_{\text{ret}} + F^{(b)}{}_{\text{adv}}]. \tag{34}$$

We also suppose the motion given to the charge to be bounded in space and time. Suppose the retarded fields of all particles in (34) are suitably attenuated at large distances from a. This implies that the advanced part of (34) also tends to zero suitably rapidly, because the only effective contribution to $\Sigma_b F^{(b)}{}_{\text{adv}}$ comes from a relatively local set of particles which have been excited by the retarded field of a. (See Figure 3.3.) Since the acceleration of a is bounded, the set of such particles tends to zero as we

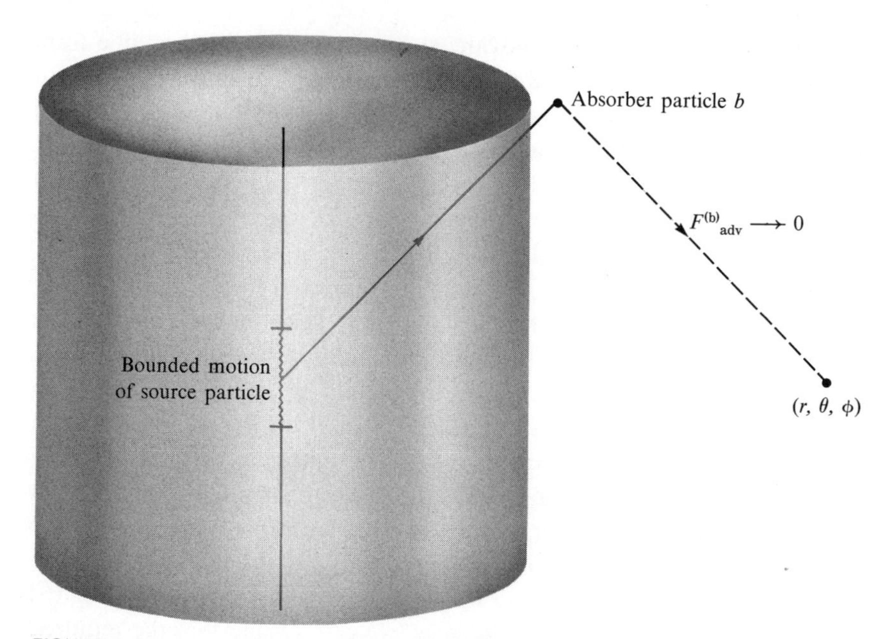

FIGURE 3.3
The retarded field due to the bounded motion of the source particle is considered to be damped by perfect absorption outside a 4-cylinder. A sufficiently large distance r from the source particle, advanced fields must come from absorber particles b that lie outside this cylinder. Hence the advanced field $F^{(b)}{}_{\text{adv}}$ is effectively zero.

evaluate $\Sigma_b\, F^{(b)}{}_{\text{adv}}$ at large distance from a. We therefore have

$$\sum_b \left[F^{(b)}{}_{\text{ret}} - F^{(b)}{}_{\text{adv}}\right] \tag{35}$$

tending to zero faster than a radiation field in vacuum. Again, since (35) satisfies a homogeneous wave equation, it must be zero everywhere. Then (34) becomes

$$F^{(a)}{}_{\text{ret}} + \sum_{b \neq a} F^{(b)}{}_{\text{ret}}, \tag{36}$$

i.e., a full retarded field.

We notice that in this derivation we give an asymmetric status to the retarded and advanced fields. The retarded fields are to be self-consistent, and hence basic. The advanced fields are secondary, arising from particles $b \neq a$ which have been excited by the retarded fields. These particles lie in the future part of the universe. We emphasize again that it does not follow that we can reverse the sign of the time coordinate, arguing that if retarded fields turn out to be self-consistent then so will advanced fields. To check the consistency of advanced solutions, a similar calculation for the *past* light cone of a has to be made, and this is not a simple reversal of the calculation for the future light cone of a.

We next examine details of such calculations in some well-known cosmological models.

Cosmological Models: Explicit Examples

We have remarked that the Robertson-Walker spaces are conformally flat. The transformation to the conformally flat form is particularly simple if $k = 0$. We shall consider models with $k = 0$ first, and in some detail. The line element

$$ds^2 = dt^2 - Q^2(t)[dr^2 + r^2(d\theta^2 + \sin^2\theta\, d\phi^2)] \tag{37}$$

transforms to

$$ds^2 = e^{2\zeta}[d\tau^2 - dr^2 - r^2(d\theta^2 + \sin^2\theta\, d\phi^2)], \tag{38}$$

with the substitutions

$$\tau = \int \frac{dt}{Q(t)}, \qquad \exp \zeta(\tau) = Q(t). \tag{39}$$

The scale of the function $Q(t)$ can be chosen for convenience, for example, by taking $Q = 1$ at the present instant. The constant of integration in (39) can also be chosen for convenience.

As we did for flat space, we Fourier-analyze the acceleration $\boldsymbol{u}(t)$ of the charge a, and consider a single component with frequency ω_0. Write

$$\boldsymbol{u} = \boldsymbol{u}_0\, e^{-i\omega_0 t}, \tag{40}$$

and choose spherical polar coordinates θ, ϕ, such that $\theta = 0$ corresponds to the direction of $\boldsymbol{u}_0$. The electric field and the magnetic field several wavelengths away from the charge are in the directions of increasing θ and ϕ, respectively. Denoting them by E_θ and H_ϕ, in flat space the retarded fields would be given in vacuum by

$$H_\phi = E_\theta = \frac{eu_0}{r} \sin\theta \exp[-i\omega_0(t - r)]. \tag{41}$$

In a conformally flat space, (41) is modified in that t is replaced by τ:

$$H_\phi = E_\theta = \frac{eu_0}{r} \sin\theta \exp[-i\omega_0(\tau - r)]. \tag{42}$$

We have to modify (42) further to include the effect of other charges, i.e., the refractive index. First we notice that, although the frequency of the wave is ω_0 with respect to the τ coordinate, proper time is given by the t coordinate. Using (39) we therefore have the proper frequency

$$\omega = \omega_0 \frac{d\tau}{dt} = \omega_0 \exp(-\zeta), \tag{43}$$

which shows that the retarded waves are redshifted, since $\exp\zeta$ increases in the future for an expanding universe.

To calculate the change of phase produced by propagation in the cosmological medium, we note that a coordinate distance dr corresponds to a proper distance

$$dz = dr \exp\zeta. \tag{44}$$

If $n - ik$ is the refractive index of the medium, the extra change of phase produced by propagation through dr is by a factor

$$\exp i(n - ik - 1)\omega\, dz = \exp i(n - ik - 1)\omega_0\, dr. \tag{45}$$

The total phase change due to propagation from a to distance r is therefore represented by the factor

$$\eta = \exp\left[\int_0^r i(n - ik - 1)\omega_0\, dr\right]. \tag{46}$$

The integral sign appears since $n - ik$ changes with r because the proper frequency of the wave depends on r, and perhaps because the density of particles in the medium changes. Thus the field at distance r is given by

$$H_\phi = E_\theta = \frac{e\, u_0 \sin\theta}{r} \exp[-i\omega_0(\tau - r)] \cdot \eta. \tag{47}$$

Here η represents the absorptive properties of the medium. For it to be a perfect absorber, we require $\eta \to 0$ with increasing r. That is, we need

$$\int_0^r k\, dr \to -\infty \tag{48}$$

as r approaches the "infinite boundary" of the universe accessible to retarded waves. This need not correspond to $r \to \infty$, as we shall see shortly.

For any cosmological model, we can calculate the integral on the lefthand side of (48). If the result turns out to tend to $-\infty$ as r increases to the "boundary" of the universe, then a self-consistent system of electrodynamics involving retarded fields is possible for such a model. If the result does not satisfy (48), then a self-consistent retarded electrodynamics is not possible for the model in question.

To calculate this integral, we need to know how k depends on ω. This dependence is determined by the physical process of absorption. We shall consider two such processes: radiative reaction, and collisional damping. Hogarth in his paper considered only the second process, and his analysis was criticized for using a process which is of thermodynamic origin. If we are to consider the problem within the framework of electrodynamics and cosmology only, we should not bring in other sources of time-asymmetry. To avoid this, we consider first the damping produced by radiative reaction.

Radiative Reaction. When the absorber particle is set in motion, it accelerates. In a universe with self-consistent retarded solutions, this acceleration produces a radiative reaction. For particle b, this is characterized by the field

$$\frac{1}{2}[F^{(b)}{}_{\text{ret}} - F^{(b)}{}_{\text{adv}}]. \tag{49}$$

If, on the other hand, we were looking for self-consistent advanced solutions, the sign of (49) would be reversed. The sign of k is thus unambiguously determined.

Suppose an electric field of magnitude $\mathcal{E}$ and frequency ω is incident on an absorber particle of mass m and charge e. The motion in the direction of the field is determined by the equation

$$m\ddot{x} = e\mathcal{E}\, e^{-i\omega t} + \frac{2e^2}{3}\dddot{x}, \tag{50}$$

so that the oscillatory displacement is

$$x = -\frac{e\mathcal{E}}{m\omega^2}\left[1 - \frac{2ie^2}{3m}\omega + O(\omega^2)\right]e^{-i\omega t}. \tag{51}$$

Taking N such charges per unit volume, the medium has a polarization Nex

in the direction of $\mathscr{E}$. Therefore the refractive index is given by $n - ik$, where

$$(n - ik)^2 \mathscr{E}\, e^{-i\omega t} = \mathscr{E}\, e^{-i\omega t} + 4\pi N e x;$$

i.e.,

$$1 - (n - ik)^2 = \frac{4\pi N e^2}{m\omega^2}\left[1 - \frac{2ie^2}{3m}\,\omega + O(\omega^2)\right]. \tag{52}$$

We have written (51) and (52) in a form suitable for low frequencies, since, as we pointed out earlier, the absorption in the future takes place at low frequencies. To solve for k from (52), we must specify the dependence of N on ω. In a Friedmann universe, we have

$$N \propto \frac{1}{Q^3}, \qquad \omega \propto \frac{1}{Q}, \qquad \text{i.e., } N \propto \omega^3. \tag{53}$$

Then we get from (52)

$$k \sim -\lambda\omega^{-2}, \qquad \lambda = \text{constant} > 0. \tag{54}$$

For the special case of the Einstein–de Sitter model, we have

$$Q(t) = \left(\frac{3Ht}{2}\right)^{2/3}, \qquad t = \frac{2}{3H} = t_0, \tag{55}$$

where H is the Hubble constant at the present epoch t_0. In conformally flat coordinates, putting τ [defined by (39)] at zero when $t = 0$, we have

$$\exp \zeta = \left(\frac{1}{2}\,H\tau\right)^2, \qquad \tau_0 = \frac{2}{H}. \tag{56}$$

A light ray emitted at the present τ_0 will reach a point with coordinate r at time $\tau = \tau_0 + r$. Hence the frequency ω is given by

$$\begin{aligned}\omega &= \omega_0 \exp -\zeta(\tau_0 + r) = \omega_0\left[\frac{1}{2}\,H(\tau_0 + r)\right]^{-2} \\ &= \omega_0\left(1 + \frac{Hr}{2}\right)^{-2}.\end{aligned} \tag{57}$$

The integral for absorption is

$$\int_0^\infty k\,dr = -\lambda\omega_0 \int_0^\infty \frac{dr}{\left(1 + \frac{Hr}{2}\right)^4}. \tag{58}$$

Since (58) converges, the absorption is incomplete, and the retarded solution is not self-consistent in this model.

We look next at the steady-state cosmological model, which is characterized by

$$N = \text{constant}, \qquad Q(t) = \exp Ht, \qquad t = 0 \text{ for present epoch,}$$

$$\exp \zeta(\tau) = (-H\tau)^{-1}, \qquad \tau = -\frac{1}{H} \text{ for present epoch.} \tag{59}$$

In this model the asymptotic form for k is

$$k \sim -\mu\omega^{-1}, \qquad \mu = \text{constant} > 0. \tag{60}$$

Arguing in the same way as above shows that at coordinate r the retarded wave has the frequency

$$\omega = \omega_0(1 - Hr), \tag{61}$$

and the absorption integral diverges:

$$\int_0^{H^{-1}} k\, dr = -\int_0^{H^{-1}} \frac{\mu}{\omega_0(1 - Hr)}\, dr = -\infty. \tag{62}$$

Thus the steady-state model has a perfect future absorber, and hence retarded solutions are self-consistent. We note in passing that the conformal transformation $t \rightarrow \tau$ limits the accessible region for retarded waves to $r \leq H^{-1}$. This is the event horizon of the steady-state universe.

Collisional Damping. In collisional damping, the absorber particle loses its energy by collisions with other particles. If ν is the collision frequency, the equation of motion of the absorber particle in the direction of the electric field $\mathscr{E}$ is given by

$$m\ddot{x} + m\nu\dot{x} = e\mathscr{E}\, e^{-i\omega t}. \tag{63}$$

We note that the collisional damping term depends on $\dot{x}$, whereas the radiative damping term depended on $\dddot{x}$. Since $\omega \rightarrow 0$ in the infinite future and $|\dot{x}|/|\dddot{x}| \propto \omega^{-2}$, the collisional damping term may well be more important than the radiative damping term, and we must therefore consider it as well.

In the limit $\omega \rightarrow 0$, we have $\omega \ll \nu$, and it becomes meaningless to seek an oscillatory solution, since the particle loses its energy by collisions before any oscillation can take place. We therefore argue in the following way.

Neglecting time-variation of $\mathscr{E}$, between collisions the particle acquires a velocity

$$\dot{x} \simeq \frac{e\mathscr{E}}{m\nu}. \tag{64}$$

The kinetic energy of N particles is therefore

$$\sim \frac{1}{2}\frac{e^2\mathscr{E}^2}{m\nu^2}N. \tag{65}$$

The electrical energy of the incident wave is $T = \mathscr{E}^2/8\pi$ per unit proper volume. Hence the attenuation is given by

$$\frac{dT}{dz} \simeq -\frac{4\pi N e^2\,T}{m\nu}, \tag{66}$$

in which N is now the particle density, and where dz is an element of proper distance directed radially outwards. To obtain (66), one simply notes that the energy (65) is abstracted from the wave in a time ν^{-1}, and that in this time the wave travels a proper distance $dz = \nu^{-1}$. Using (44), we get

$$\frac{dT}{dr} \simeq -\frac{4\pi N e^2}{m\nu}\exp\zeta \cdot T. \tag{67}$$

The absorption integral is therefore of the form

$$-\int \frac{4\pi N e^2}{m\nu} e^{\zeta}\, dr. \tag{68}$$

For the steady state model, $N =$ constant, $\nu =$ constant, and the integral (68) is similar to (62) and therefore leads to the same conclusion. For the Einstein–de Sitter model, ν may well behave in a complicated way because the velocities of the particles in the medium are dependent on r. However, the proportionality

$$\nu \propto N^{-1} \tag{69}$$

introduces strong convergence in (68), since $N \propto e^{-3\zeta}$. Hence we expect (68) to remain convergent, and for the previous conclusions to remain unaltered. This question is further discussed in Chapter 9.

As in the retarded case, we can do similar calculations for the advanced solutions. The results are summarized in Table 3.1. We should point out, however, that these calculations are based on an extrapolation of classical physics to $\omega \to \infty$, with the added assumption that absorption cross sections saturate at finite values. These calculations are not as soundly based as those for retarded waves, since they ignore quantum effects, which are likely to be important.

We can also perform similar calculations in the $k = \pm 1$ Robertson-Walker spaces, since they are also conformally flat. Such calculations have been made by Roe, and his conclusions are included in Table 3.1.

TABLE 3.1
Absorber properties of cosmological models.

Cosmological model	*Future absorber*	*Past absorber*	*Electromagnetic propagation*
Minkowski	Perfect	Perfect	Ambiguous
Friedmann ($k = 0$, $k = -1$)	Imperfect	Perfect	Advanced
Friedmann ($k = +1$)	Perfect	Perfect	Ambiguous
Steady state	Perfect	Imperfect	Retarded

§3.4. CONCLUSIONS AND RESTATEMENT OF PROCEDURE

We have just seen that the Friedmann models with $k = 0, -1$, do not give a perfect absorber along the future light cone, and hence are unsuitable for obtaining the empirically observed electromagnetic arrow of time within the framework of the Fokker action and of the Wheeler-Feynman theory. Moreover, models with $k = +1$ lead to the same ambiguous situation as did a static universe. It follows that, unless we abandon the present line of development and return to the unsatisfactory *ad hoc* choice of retarded potentials, and the unsatisfactory position concerning the self-force on a charged particle, we must discard the Friedmann models.

The steady-state model, on the other hand, meets the necessary conditions. Retarded potentials are self-consistent but advanced potentials are not; so this model is free from ambiguity. However, we cannot conclude from this success that the steady-state theory is correct, because there may be other models, not considered above, which also meet the necessary conditions. We shall in fact encounter another such model, different from any which has hitherto been considered by cosmologists, in Chapter 9.

It will be useful by way of concluding this chapter to restate the procedure which has been followed. Now that the details of the various calculations are out of the way, it is possible to see the basic ideas more clearly. The electromagnetic field is formally symmetric with respect to advanced and retarded fields, and is given by

$$\sum_a [F^{(a)}{}_{\text{ret}} + F^{(a)}{}_{\text{adv}}]. \tag{70}$$

If the particle paths were known, (70) could be calculated from (3) and (4). But the particle paths cannot be specified independently of each other, since they are interrelated through (10). Because we have no *a priori* knowledge of these interrelations, we are in the awkward situation that until we know the paths we cannot calculate the fields of the particles, and until we know the fields we cannot calculate the paths. To break this deadlock, we make an assumption designed to permit the interrelations of the particles to be disentangled. We then proceed to calculate the paths and the fields, and we check to see whether the result is consistent with the initial assumption. If it is, a self-consistent cycle of argument has been established—we have a possible way in which the universe can behave. If several such cycles can be found, then we have several possible modes of behavior for the universe. In such a situation, we would have no way of discovering which mode our universe should adopt—there would be ambiguity. But if we should find one consistent cycle of argument, and no others, we should have determined the only possible behavior of the universe. This is the situation we hope to arrive at.

We have been especially interested in an initial assumption which states: "The interrelation of particles $b \neq a$ with the motion of any particle a is to be calculated using the fully retarded field of a." We hope to find this assumption self-consistent, because it agrees with experience. And we hope to find the corresponding assumption for advanced fields to be inconsistent. This distinction between retarded and advanced fields turns out to be related to the expansion of the universe and to the cosmological model. Hence the theoretical development is capable not only of resolving unsatisfactory features of Maxwell's theory, but also of giving information which narrows the range of acceptable cosmological models.

4

THE QUANTUM RESPONSE OF THE UNIVERSE

§4.1. QUANTIZATION OF DIRECT-PARTICLE THEORIES (LECTURE 7)

We have so far proceeded along classical lines. We have shown that the direct-particle approach to electromagnetism works at least as well as the Maxwell field approach in explaining all the classical phenomena of the interaction of charges, and that it links with cosmology in an interesting way. The choice of retarded potentials is not an *ad hoc* choice, but is dictated by the universe in the large. Moreover, the unbounded motions of charges moving under the self-force do not arise in this theory.

Although success in the classical domain is necessary for any physical theory, it is not sufficient. Nature, as we understand it today, is quantum in character. In electrodynamics, quantum theory has unearthed a vast collection of phenomena outside the concepts of classical physics. These have been explained with remarkable success by the quantized version of Maxwell's theory, although there have been conceptual and mathematical stumbling blocks too. Can the direct-particle theory do as well here, if not better?

At first sight, an attempt to extend the classical direct-particle theory to include quantum phenomena seems unlikely to succeed. In Maxwell's theory we have fields to quantize. The degrees of freedom of these fields result in packets of energy called "photons," which play such an important part in quantum electrodynamics. We have no analogous degrees of freedom in direct-particle fields. Thus photons do not appear to exist in the latter theory. The only degrees of freedom are those vested in the particles. Can we get all the conventional quantum electrodynamics by a first quantization alone? If not, the theory fails. If we can, however, the theory must be regarded as the superior theory, because it reproduces all observations under fewer degrees of freedom.

Other difficulties can be anticipated concerning particles and antiparticles. In classical theory all world lines are endless and timelike. In relativistic quantum electrodynamics, the world lines can go forward and backward in time. What happens to the "identity" of a world line under these circumstances? How does the rule of no self-interaction operate under such conditions?

These are some of the problems which arise when we undertake to quantize the Fokker theory. We shall proceed by stages in solving them, beginning with the simpler nonrelativistic picture and ending with fully relativistic interactions of electrons and positrons.

§4.2. THE PATH-INTEGRAL APPROACH TO QUANTUM MECHANICS

As we said, the only quantities which can be quantized in the direct-particle theory are the particle motions. The usual procedure is to start with a classical Hamiltonian and introduce operators. This procedure is not helpful here, since we have no Hamiltonian to start with. We have the action, the Fokker action, to describe the classical theory. The most suitable method of quantization is the path-integral approach developed by Feynman in 1948. This approach has the advantage of using the classical action as its starting point, although it has so far been developed for nonrelativistic particles only. This restriction we shall remove later.

The details of the method of path integrals can be found in Feynman and Hibbs (1965). Since the techniques are still not widely known, we shall go briefly over some of the main ideas.

Path Amplitudes

Suppose a physical system has action S. This is defined in terms of "paths" Γ that the system can follow in coordinate space. Classical physics tells us that not all Γ are permissible. In general, there is a unique path Γ from a given point P_1 in coordinate space to a given point P_2. Writing Γ_0 for this path, Γ_0 is given by the principle of stationary action

$$\delta S = 0 \qquad \text{for } \Gamma = \Gamma_0. \tag{1}$$

Earlier, we have used this mysterious prescription, and have noted its remarkable success in classical physics. In 1938, Dirac suggested the interesting idea, later developed quantitatively by Feynman, that (1) is the consequence of a more general principle operating in quantum mechanics. In quantum theory the system is permitted any of the paths Γ from P_1 to P_2, but each path has a definite probability amplitude proportional to

$$\exp(iS/\hbar), \tag{2}$$

where $\hbar$ is Planck's constant. All amplitudes add, giving a total amplitude for the system to go from P_1 to P_2. In the classical limit $\hbar \to 0$, and (2) oscillates wildly as we move from path to path—with the exception of Γ_0 where (1) holds. Paths in the neighborhood of Γ_0 make a significant contribution in this limit, whereas the contributions from other paths average out to zero. Hence the classical principle of stationary action.

Feynman carried these ideas further by introducing the concept of the path integral. He defined the nonrelativistic quantum-mechanical propagator $K(P_2; P_1)$ for the system to go from P_1 to P_2 by

$$\begin{aligned} K(P_2; P_1) &= \sum_{\Gamma} (\text{constant}) \exp(iS/\hbar), \qquad && t_2 > t_1, \\ &= 0, && t_2 < t_1. \end{aligned} \tag{3}$$

In (3) the action S is computed for each path Γ, and the sum is over all Γ from P_1 to P_2. The constant is a normalization constant. If, as is usual, the paths form a continuum, the sum is replaced by an integral

$$K(P_2; P_1) = \int \exp[iS(\Gamma)/\hbar]\mathcal{D}\Gamma, \qquad t_2 > t_1. \tag{4}$$

The integral is over the continuum of paths and is more complicated than the Riemann or the Lebesgue integral, which are summed over sets of points. Only limited progress has been made toward giving a rigorous mathematical foundation to this concept. Feynman was able, however, to obtain all required physical answers by various subtle devices. We shall draw heavily upon these techniques in the subsequent work. The constant in (3) can be absorbed in the measure of $\mathscr{D}\Gamma$.

Suppose P_1 represents the spacetime point $(\boldsymbol{x}_1, t_1)$ and P_2 the point $(\boldsymbol{x}_2, t_2)$ in the motion of a particle. Let $t_2 > t_1$. Any path Γ from P_1 to P_2 will pass through some intermediate point P having time coordinate t. Let the spatial position of P be $\boldsymbol{x}$. Since the action functional is additive over paths, we can write

$$S(\Gamma_{P_2P_1}) = S(\Gamma_{P_2P}) + S(\Gamma_{PP_1}), \tag{5}$$

where the path $\Gamma_{P_2P_1}$ from P_1 to P_2 is made up of the segment Γ_{PP_1} from P_1 to P and the segment Γ_{P_2P} from P to P_2. Hence

$$\exp[iS(\Gamma_{P_2P_1})/\hbar] = \exp[iS(\Gamma_{P_2P})/\hbar] \exp[iS(\Gamma_{PP_1}/\hbar]. \tag{6}$$

The sum over all paths from P_1 to P_2 can be obtained by summing all paths Γ_{PP_1} and Γ_{P_2P} and integrating over the spatial coordinates of P. From (6), together with an appropriate measure for the path integral (4), we get

$$K(\boldsymbol{x}_2, t_2; \boldsymbol{x}_1, t_1) = \int K(\boldsymbol{x}_2, t_2; \boldsymbol{x}, t)\, K(\boldsymbol{x}, t; \boldsymbol{x}_1, t_1)\, d^3\boldsymbol{x}. \tag{7}$$

Together, (4) and (7) suggest an alternative way of defining the path amplitude. Suppose we divide $\Gamma_{P_2P_1}$ into a large number of small segments with intermediate points $Q_r (r = 1, \ldots, N-1)$. We can define $Q_0 = P_1$ and $Q_N = P_2$. Over each segment (Q_{r-1}, Q_r) we may imagine S to change very slowly. We define the amplitude for such a segment to be proportional to $K(Q_r; Q_{r-1})$. The amplitude along the entire path is then given by the product

$$\lim_{N\to\infty} \prod_{r=1}^{N} A_r K(Q_r; Q_{r-1}), \tag{8}$$

where A_r is a constant of proportionality with the dimensions of spatial volume. Summing over all paths gives

$$\int \prod_{r=1}^{N} A_r K(Q_r; Q_{r-1}) \mathscr{D}\Gamma = \int \ldots \int K(P_2; Q_{N-1})\, K(Q_{N-1}; Q_{N-2}) \ldots$$

$$\ldots K(Q_2; Q_1)\, K(Q_1; P_1)\, d^3Q_{N-1} \ldots d^3Q_1 = K(P_2; P_1) \tag{9}$$

by (7), the integrations of d^3Q_r being over the time sections $t = t_r$; $r = 1, \ldots, N-1$.

The constants A_r are again absorbed in the measure of Γ. Because of (7), the definition (8) leads to the same function $K(P_2; P_1)$ as before.

Sometimes we know K but not S. Then (8) is useful to define a path amplitude. The original definition (2) is the more direct one, however. We shall use (8) in relativistic path-integral theory.

The Wavefunction.

Suppose that, instead of knowing that the particle is at $(\boldsymbol{x}_1, t_1)$, we only know the probability amplitude $\psi(\boldsymbol{x}_1, t_1)$ for it to be at various spatial positions $\boldsymbol{x}_1$ on the time section $t = t_1$. We then ask, what is the probability amplitude $\phi(\boldsymbol{x}_2, t_2)$ of finding the particle at $(\boldsymbol{x}_2, t_2)$? This is obtained by the weighted mean of $K(\boldsymbol{x}_2, t_2; \boldsymbol{x}_1, t_1)$,

$$\phi(\boldsymbol{x}_2, t_2) = \int K(\boldsymbol{x}_2, t_2; \boldsymbol{x}_1, t_1)\, \psi(\boldsymbol{x}_1, t_1)\, d^3\boldsymbol{x}_1. \tag{10}$$

As $t_2 \to t_1$, $\phi \to \psi$. This implies

$$\lim_{t_2 \to t_1} K(\boldsymbol{x}_2, t_2; \boldsymbol{x}_1, t_1) = \delta_3(\boldsymbol{x}_2 - \boldsymbol{x}_1). \tag{11}$$

The function ϕ is the usual Schrödinger wavefunction. Does it satisfy the Schrödinger equation? The answer is "yes," and we illustrate this by a simple example. For a particle moving in a potential field V, we have

$$S = \int \left(\frac{1}{2} m\dot{x}^2 - V\right) dt. \tag{12}$$

If we substitute this in (4) and use (10), we find that ϕ satisfies the differential equation

$$-\frac{\hbar^2}{2m} \nabla_2{}^2\phi + V\phi = i\hbar \frac{\partial \phi}{\partial t_2}. \tag{13}$$

This remarkable result is proved in Appendix 2, p. 216. In view of (11), and the fact that $K = 0$ for $t_2 < t_1$ (no propagation backwards in time), (13) implies that K satisfies the equation

$$\left(\frac{\partial}{\partial t_2} - \frac{i\hbar}{2m} \nabla_2{}^2 + \frac{i}{\hbar} V\right) K(\boldsymbol{x}_2, t_2; \boldsymbol{x}_1, t_1) = \delta_3(\boldsymbol{x}_2 - \boldsymbol{x}_1)\, \delta(t_2 - t_1). \tag{14}$$

Transition Probability

Let the particle be in an initial state $\phi_i(\boldsymbol{x}, t_1)$. We wish to determine the probability that it is in state $\phi_f(\boldsymbol{x}, t_2)$ for $t_2 > t_1$. Using (10), the amplitude is given by

$$
\begin{aligned}
\langle\phi_f|\phi_i\rangle &= \int\int \phi_f^*(\boldsymbol{x}_2, t_2)\, K(\boldsymbol{x}_2, t_2; \boldsymbol{x}_1, t_1)\, \phi_i(\boldsymbol{x}_1, t_1)\, d^3\boldsymbol{x}_2\, d^3\boldsymbol{x}_1 \\
&= \int\int\int \phi_f^*(\boldsymbol{x}_2, t_2) \exp[iS(\Gamma_{21})/\hbar]\, \phi_i(\boldsymbol{x}_1, t_1) \mathcal{D}\Gamma_{21}\, d^3\boldsymbol{x}_2\, d^3\boldsymbol{x}_1, \qquad (15)
\end{aligned}
$$

where * denotes complex conjugate.

The transition probability from ϕ_i to ϕ_f is given by the square of the modulus of (15), i.e., by

$$
\begin{aligned}
P(\phi_i \rightarrow \phi_f) &= |\langle\phi_f|\phi_i\rangle|^2 \\
&= \int\int\int\int\int\int \phi_f^*(\boldsymbol{x}_2, t_2) \exp[iS(\Gamma_{21})/\hbar]\, \phi_i(\boldsymbol{x}_1, t_1) \\
&\qquad \cdot \phi_i^*(\boldsymbol{x}'_1, t_1) \exp[-iS(\Gamma'_{21}/\hbar)] \phi_f(\boldsymbol{x}'_2, t_2) \\
&\qquad \cdot \mathcal{D}\Gamma_{21} \mathcal{D}\Gamma'_{21}\, d^3\boldsymbol{x}_1\, d^3\boldsymbol{x}_2\, d^3\boldsymbol{x}'_1\, d^3\boldsymbol{x}'_2. \qquad (16)
\end{aligned}
$$

We shall often refer to Γ'_{21} as a conjugate path, implying that the action along Γ'_{21} is multiplied by $-i$ instead of $+i$.

Perturbation Theory

We now consider the problem of perturbation expansion from the path-integral point of view. Suppose we know the behavior of a system with action S_0. Let the system be disturbed by a potential $V(\boldsymbol{x}, t)$ over the time interval (t_1, t_2). The action during this interval is therefore

$$S = S_0 - \int_{t_1}^{t_2} V(\boldsymbol{x}, t)\, dt. \qquad (17)$$

We now wish to determine $K_V(\boldsymbol{x}_2, t_2; \boldsymbol{x}_1, t_1)$ in terms of the unperturbed propagator $K_0(\boldsymbol{x}_2, t_2; \boldsymbol{x}_1, t_1)$. We have

$$K_0(\boldsymbol{x}_2, t_2; \boldsymbol{x}_1, t_1) = \int \exp\left[\frac{iS_0(\Gamma_{21})}{\hbar}\right] \mathcal{D}\Gamma_{21}, \qquad (18)$$

$$K_V(\boldsymbol{x}_2, t_2; \boldsymbol{x}_1, t_1) = \int \exp\left[\frac{iS_0(\Gamma_{21})}{\hbar} - \frac{i}{\hbar}\int_{\Gamma_{21}} V\, dt\right] \mathcal{D}\Gamma_{21}. \qquad (19)$$

Expanding the exponential involving the integral over V, then

$$K_V(\boldsymbol{x}_2, t_2; \boldsymbol{x}_1, t_1) = \int \exp\left[\frac{iS_0(\Gamma_{21})}{\hbar}\right] \cdot \left[1 - \frac{i}{\hbar}\int_{\Gamma_{21}} V\,dt - \frac{1}{2\hbar^2}\left(\int_{\Gamma_{21}} V\,dt\right)^2 + \cdots\right] \mathscr{D}\Gamma_{21}. \quad (20)$$

The path integral for the unity term in this expansion gives K_0. We consider the next term

$$-\int \exp\left[\frac{iS_0(\Gamma_{21})}{\hbar}\right] \cdot \frac{i}{\hbar}\int_{\Gamma_{21}} V\,dt\,\mathscr{D}\Gamma_{21}. \quad (21)$$

The integral $\int_{\Gamma_{21}} V\,dt$ is over a specific path Γ_{21}, given by a function $\boldsymbol{x}(t)$. Suppose we take a particular instant t of time and take all paths which pass through $\boldsymbol{x}(t)$ on their way from P_1 to P_2. If we sum over these paths alone, we will get, as in the analysis leading to (7), the product

$$K_0(\boldsymbol{x}_2, t_2; \boldsymbol{x}, t)K_0(\boldsymbol{x}, t; \boldsymbol{x}_1, t_1). \quad (22)$$

If the integral over V were absent, we would simply have got (7). But now we have to weight (22) with $-\dfrac{iV(\boldsymbol{x}, t)}{\hbar}$. Then we have to sum over all $\boldsymbol{x}$, to include all paths from 1 to 2. Finally we perform the time-integral to get the entire contribution:

$$-\frac{i}{\hbar}\int_{t_1}^{t_2} K_0(\boldsymbol{x}_2, t_2; \boldsymbol{x}, t)V(\boldsymbol{x}, t)K_0(\boldsymbol{x}, t; \boldsymbol{x}_1, t_1)\,d^3\boldsymbol{x}\,dt. \quad (23)$$

The physical meaning of this operation is illustrated in Figure 4.1. If we consider a given t and $\boldsymbol{x}$, we imagine the system to proceed undisturbed from $(\boldsymbol{x}_1, t_1)$ to $(\boldsymbol{x}, t)$. Then it is scattered by $V(\boldsymbol{x}, t)$, after which it proceeds undisturbed from $(\boldsymbol{x}, t)$ to $(\boldsymbol{x}_2, t_2)$. This scattering could occur anywhere within the spacetime slab $t_1 \leq t \leq t_2$. Hence the integration in (23). The four-dimensional volume element, $d^3\boldsymbol{x}\,dt = d\tau$, say, can actually be taken over the whole of spacetime, since one of the K_0 functions involves propagation backward in time. The integrand is therefore zero whenever the point $\boldsymbol{x}$, t, falls outside the slab $t_1 \leq t \leq t_2$.

Higher-order terms in the expansion give multiple scattering processes. Thus (19) is the closed form of the usual infinite perturbation series:

$$K_V(\boldsymbol{x}_2, t_2; \boldsymbol{x}_1, t_1) = K_0(\boldsymbol{x}_2, t_2; \boldsymbol{x}_1, t_1)$$

$$- \frac{i}{\hbar} \int K_0(\boldsymbol{x}_2, t_2; \boldsymbol{x}, t) V(\boldsymbol{x}, t) K_0(\boldsymbol{x}, t; \boldsymbol{x}_1, t_1)\, d\tau$$

$$+ \left(\frac{i}{\hbar}\right)^2 \int\int K_0(\boldsymbol{x}_2, t_2; \boldsymbol{x}_3, t_3) V(\boldsymbol{x}_3, t_3) K_0(\boldsymbol{x}_3, t_3; \boldsymbol{x}_4, t_4) V(\boldsymbol{x}_4, t_4)$$

$$\cdot K_0(\boldsymbol{x}_4, t_4; \boldsymbol{x}_1, t_1)\, d\tau_3\, d\tau_4 + \dots . \tag{24}$$

Transition Element

Classically we are used to continuous changes of dynamical variables. In quantum mechanics, transitions are in general discontinuous. Can we give meaning to "velocity" or "acceleration" in discontinuous transitions? In the path-integral formulation, we can indeed give a meaning to such terms, by means of the concept of transition elements. Suppose ϕ_i and ϕ_f are the initial and final states of a system described by action S. We have already seen that the probability amplitude can be defined by the path integral

$$\langle \phi_f | \phi_i \rangle = \int\int\int \phi_f^* \exp(iS/\hbar) \phi_i \mathscr{D}\Gamma\, d^3\boldsymbol{x}_1\, d^3\boldsymbol{x}_2. \tag{25}$$

Suppose we now have a functional $F[\Gamma]$ of a path Γ. Then the transition

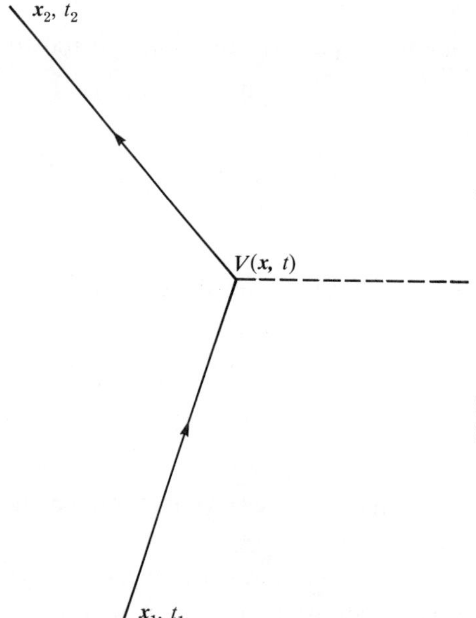

FIGURE 4.1
Scattering due to a potential field $V(\boldsymbol{x}, t)$.

element of F is given by

$$\langle\phi_f|F|\phi_i\rangle = \iiint \phi_f^* \exp\left[\frac{iS(\Gamma)}{\hbar}\right] F[\Gamma]\phi_i \mathscr{D}\Gamma\, d^3\mathbf{x}_1\, d^3\mathbf{x}_2. \tag{26}$$

This definition means that $\langle\phi_f|F|\phi_i\rangle$ is a certain kind of average over all paths suitably weighted by the initial and final wavefunctions. Classically, we could have calculated $F[\Gamma]$ exactly, since we know that $\Gamma = \Gamma_0$ is the unique solution. In quantum mechanics, we cannot make such a definitive statement.

Equation (26) is easy to apply in simple cases. For instance, for a free particle of mass m, the transition element of velocity is given by

$$\langle\phi_f|\dot{x}|\phi_i\rangle = \int -\frac{i\hbar}{m}\phi_f^* \nabla_x\phi_i\, d^3\mathbf{x}. \tag{27}$$

As seen in this example, a transition element need not be real even if the original dynamical variable is real.

Influence Functional

We come now to the concept most useful for the quantum development of direct-particle theories. Suppose we have one quantum-mechanical system in interaction with another, and suppose we are interested only in the detailed behavior of the first system regardless of what happens to the second. We may speak of the second system as the "external environment" of the first. To fix ideas, let the first system be described by a coordinate q and the second by Q. The combined action for the two systems is taken to be of the form

$$S = S_0[q(t)] + S_E[Q(t)] + S_I[q(t), Q(t)], \tag{28}$$

where S_0 represents the action of the first system alone, S_E is the action of the second system alone, and S_I represents the interaction of the two systems. Let $\phi_i(q_i, t_i)$ be the initial state and $\phi_f(q_f, t_f)$ the final state of the first system. Using (16), the probability of transition $\phi_i \to \phi_f$ is given by

$$P(\phi_i \to \phi_f) = \iiiiii \phi_f^*(q_f, t_f)\phi_i^*(q_i', t_i)\phi_i(q_i, t_i)\phi_f(q_f', t_f)$$

$$\cdot \exp\left(\frac{i}{\hbar}\{S_0[q(t)] - S_0[q'(t)]\}\right) F[q, q']\mathscr{D}q\mathscr{D}q'\, dq_f\, dq_f'\, dq_i\, dq_i', \tag{29}$$

where

$$F[q(t), q'(t)] = \sum_f \langle\psi_f\left|\exp\left\{\frac{i}{\hbar} S_I[q(t), Q(t)]\right\}\right|\psi_i\rangle \cdot \langle\psi_f\left|\exp\left\{\frac{i}{\hbar} S_I[q'(t), Q'(t)]\right\}\right|\psi_i\rangle^*. \quad (30)$$

Here ψ_i is the initial state of the second system. We have summed over all final states ψ_f, since we are not interested in how the environment ends up.

$F[q(t), q'(t)]$ is called the influence functional and represents the "force" exerted by the environment on the first system.

Feynman and Hibbs discuss influence functionals and their applications in some detail; so we shall not discuss them further. We expect the universe to act as the external environment in the quantum version of the Wheeler-Feynman theory. It is our aim to determine the quantum analogue of the response of the universe that earlier we calculated classically. This response will appear in the form of an influence functional.

§4.3. ATOMIC TRANSITIONS (LECTURE 8)

We now discuss how the Wheeler-Feynman theory operates in nonrelativistic quantum mechanics by considering a typical example—that of atomic transitions. Consider an atomic electron making transitions between states of energy E_n and E_m. If $E_m > E_n$, we know that the electron jumps "down" from the state of energy E_m to the state of energy E_n even if no external field is present. Such jumps are called spontaneous transitions to distinguish them from induced transitions caused by ambient electromagnetic fields. The latter are symmetric with respect to any pair of states; i.e., the probability of jumping down is the same as that for jumping up. In Maxwell field theory, the spontaneous transitions are explained by means of field quantization in which fluctuations of the vacuum (no field) state play a crucial part. It would seem, therefore, that here at last we have a phenomenon which really depends on the "field" character of electromagnetism. How can the direct-particle theory explain the downward transition of an electron when apparently there is no electromagnetic disturbance present?

The clue to the answer to this question is found in the classical formula for a universe that is completely absorbing along the future light cone. We have

$$\sum_{b \neq a} \frac{1}{2} [F_{\text{adv}}^{(b)} + F_{\text{ret}}^{(b)}] = \sum_{b \neq a} F_{\text{ret}}^{(b)} + \frac{1}{2} [F_{\text{ret}}^{(a)} - F_{\text{adv}}^{(a)}]. \tag{31}$$

In (31) the first term on the righthand side shows how other charges b influence the motion of the charge a via retarded fields. This term accounts for the induced transitions of a. The usual results for upward and downward transitions can be recovered from it. The second term, on the other hand, appears to concern charge a alone, and would be present even if the first term were zero. It is to this second term that we must look for spontaneous transitions.

We have seen how the second term arises. It is the response of the rest of the universe to the retarded field of a. This response arises instantaneously with the outward movement of a disturbance from a. In the quantum theory of atomic transitions, we expect the following chain of events: an electron a initially in an upper-energy state could stay there, or it could jump up or jump down. Quantum-mechanically, we must consider all possible motions of a, not just a unique classical one. Each motion would invoke a response analogous to (31). We then ask whether this response is consistent with the over-all transition of a, and if it is, we calculate the probability of transition. It is the nature of the response from the universe which will decide whether there should be transition at all, and, if so, with what probability. Thus the apparently local process of spontaneous transition will turn out to depend on the large-scale structure of the universe.

To fix ideas, let us consider the motion of a in the time-interval $0 \leq t \leq T$ (see Figure 4.2), and denote the displacement of a by $\boldsymbol{a}(t)$. Let b be a typical absorber particle whose world line is intersected in intervals Δ_- and Δ_+, respectively, by the past and future light cones from the initial point $[\boldsymbol{a}(0), 0]$ and the final point $[\boldsymbol{a}(T), T]$ on the path $\boldsymbol{a}(t)$ of a. From what has been said above, the induced transitions of a arise from its interaction with the retarded field of b, i.e., from the portion Δ_- of the world line of b. The action governing induced transitions is therefore

$$-e_a \int_0^T \boldsymbol{A}_{\text{ret}}^{(b)}(\boldsymbol{a}) \cdot \dot{\boldsymbol{a}}\, dt, \tag{32}$$

where $\boldsymbol{A}_{\text{ret}}^{(b)}$ is the full retarded 3-potential from b. To calculate spontaneous transitions, we need, on the other hand, the transitions of b induced by the full retarded field of a. The action governing this is

$$-e_b \int_{\Delta_+} \boldsymbol{A}_{\text{ret}}^{(a)}(\boldsymbol{b}) \cdot \dot{\boldsymbol{b}}\, dt, \tag{33}$$

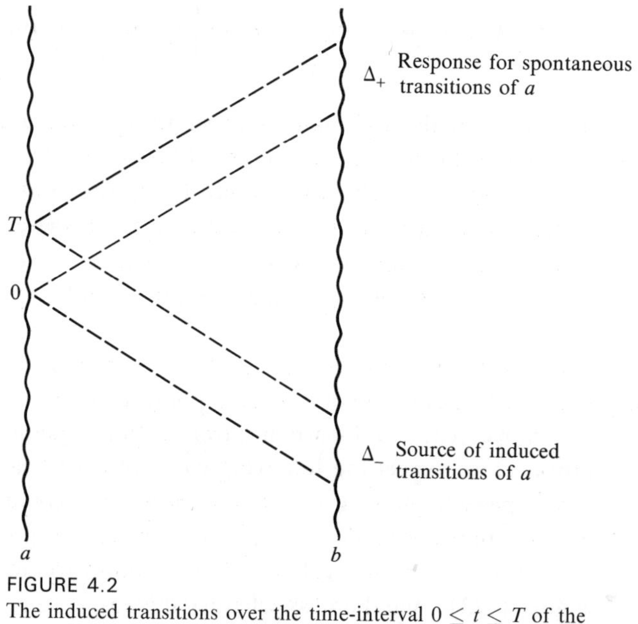

FIGURE 4.2
The induced transitions over the time-interval $0 \leq t \leq T$ of the particle a arise from interaction with the past segment Δ_- of the path of particle b, whereas spontaneous transitions of a arise from interaction with the future segment Δ_+.

with a similar notation. As in the classical case, we are looking for a self-consistent cycle of argument in which the net field is the retarded one. Throughout this calculation, we shall work in a conformally flat cosmological model with a perfect future absorber.

Our aim is to calculate the influence functional governing the motion of a and arising from the whole universe. Since we may assume the different absorber particles to act independently, the influence functional has the form

$$F[\boldsymbol{a}(t), \boldsymbol{a}'(t)] = \prod_{b \neq a} F^{(b)}[\boldsymbol{a}(t), \boldsymbol{a}'(t)], \tag{34}$$

where $F^{(b)}[\boldsymbol{a}(t), \boldsymbol{a}'(t)]$ is the influence functional exerted by a typical particle b.

To calculate $F^{(b)}[\boldsymbol{a}(t), \boldsymbol{a}'(t)]$, we consider all transitions produced by (33) in the absorber particle b. If $\psi_i(\boldsymbol{b})$ is the initial wavefunction and $\psi_f(\boldsymbol{b})$ any final wavefunction of b, we have from (30)

$$F^{(b)}[\boldsymbol{a}(t), \boldsymbol{a}'(t)]$$

$$= \sum_f \int\int\int\int \psi_f^*(\boldsymbol{b}_f)\psi_f(\boldsymbol{b}'_f)J^{(b)}\psi_i(\boldsymbol{b}_i)\psi_i^*(\boldsymbol{b}'_i)\, d^3\boldsymbol{b}_i\, d^3\boldsymbol{b}_f\, d^3\boldsymbol{b}'_i\, d^3\boldsymbol{b}'_f, \quad (35)$$

where

$$J^{(b)} = \int\int \exp\frac{i}{\hbar}\{S_E[\boldsymbol{b}(t)] - S_E[\boldsymbol{b}'(t)] + S_I[\boldsymbol{a}(t), \boldsymbol{b}(t)] - S_I[\boldsymbol{a}'(t), \boldsymbol{b}'(t)]\} \cdot \mathcal{D}\boldsymbol{b}\mathcal{D}\boldsymbol{b}'. \quad (36)$$

With the interaction governed by (33), we write

$$S_I[\boldsymbol{a}(t), \boldsymbol{b}(t)] = -e_b \int_{\Delta_+} \boldsymbol{A}_{\text{ret}}^{(a)}(\boldsymbol{b}) \cdot \dot{\boldsymbol{b}}\, dt \quad (37)$$

in (36), with a similar expression for $S_I[\boldsymbol{a}'(t), \boldsymbol{b}'(t)]$.

Because of the cooling produced by the expansion of the universe, b is usually in its ground state. We shall take ψ_i to be the ground state. We can expand the part of the exponential in (36) containing the interaction terms only. The unity term in the expansion corresponds to $\psi_f \rightarrow \psi_i$. Thus we have

$$F^{(b)}[\boldsymbol{a}, \boldsymbol{a}'] = 1 + \sum_{f\neq i} \int\int\int\int\int\int \psi_f^*(\boldsymbol{b}_f)\psi_f(\boldsymbol{b}'_f)\psi_i^*(\boldsymbol{b}'_i)\psi_i(\boldsymbol{b}_i)$$

$$\cdot \frac{1}{\hbar^2}\left[e_b^2 \int_{\Delta_+} \boldsymbol{A}_{\text{ret}}^{(a)}(\boldsymbol{b}) \cdot \dot{\boldsymbol{b}}\, dt \int_{\Delta'_+} \boldsymbol{A}_{\text{ret}}^{(a')}(\boldsymbol{b}') \cdot \dot{\boldsymbol{b}}'\, dt\right]$$

$$\cdot \exp\left(\frac{i}{\hbar}\{S_E[\boldsymbol{b}(t)] - S_E[\boldsymbol{b}'(t)]\}\right)\mathcal{D}\boldsymbol{b}'\mathcal{D}\boldsymbol{b}\, d^3\boldsymbol{b}_f\, d^3\boldsymbol{b}_i\, d^3\boldsymbol{b}'_i\, d^3\boldsymbol{b}'_f$$

$$+ \text{ terms in } \left[\int_{\Delta_+}\right]^2, \left[\int_{\Delta'_+}\right]^2 + \dots, \quad (38)$$

where $\boldsymbol{A}_{\text{ret}}^{(a)}$ is calculated for the path $\boldsymbol{a}(t)$, $\boldsymbol{A}_{\text{ret}}^{(a')}$ for the conjugate path $\boldsymbol{a}'(t)$.

Let E_f and E_i be the energies of states ψ_f and ψ_i, respectively. Then first-order perturbation theory shows that the contribution to the above expression from the transition $\psi_i \rightarrow \psi_f$ is

$$\frac{e_b^2}{\hbar^2}\left(\frac{E_f - E_i}{\hbar}\right)^2 M[\boldsymbol{a}(t)]M^*[\boldsymbol{a}'(t)], \tag{39}$$

where

$$M[\boldsymbol{a}(t)] = \int_{\Delta_+} \exp\left[\frac{i(E_f - E_i)t}{\hbar}\right] dt \int \psi_f^*(\boldsymbol{b}) \boldsymbol{A}_{\text{ret}}^{(a)}(\boldsymbol{b}) \cdot \boldsymbol{b}\psi_i(\boldsymbol{b})\, d^3\boldsymbol{b}. \tag{40}$$

To calculate (40), we proceed as follows. Take an origin near a to measure the displacement $\boldsymbol{a}(t)$. Similarly, let $\boldsymbol{b}(t)$ measure the displacement from an origin near b. Let the origin near b have a relative displacement $\boldsymbol{R}$ with respect to the origin near a. Then the vector from $\boldsymbol{a}(t)$ to $\boldsymbol{b}(t)$ is given by

$$\boldsymbol{r} = \boldsymbol{R} + \boldsymbol{b} - \boldsymbol{a}. \tag{41}$$

For $|\boldsymbol{R}| \gg |\boldsymbol{b}|, |\boldsymbol{a}|$, we get

$$r = |\boldsymbol{r}| \cong R + \frac{1}{R}\boldsymbol{R} \cdot (\boldsymbol{b} - \boldsymbol{a}). \tag{42}$$

In the absence of any dispersion, we would have, in the Coulomb gauge,

$$\boldsymbol{A}_{\text{ret}}^{(a)}(\boldsymbol{b}) = \frac{e_a}{r - \boldsymbol{r} \cdot \dot{\boldsymbol{a}}} \sum_{j=1,2} [\boldsymbol{\alpha}^{(j)} \cdot \dot{\boldsymbol{a}}]\boldsymbol{\alpha}^{(j)}, \tag{43}$$

where $\boldsymbol{\alpha}^{(j)}$ are two unit vectors which form a mutually orthogonal triad with $\boldsymbol{r}/|\boldsymbol{r}|$.

The quantity $r - \boldsymbol{r} \cdot \dot{\boldsymbol{a}}$ varies only slightly with $\boldsymbol{b}$, and it is sufficiently accurate to replace it in (43) by $R - \boldsymbol{R} \cdot \dot{\boldsymbol{a}}$,

$$\boldsymbol{A}_{\text{ret}}^{(a)}(\boldsymbol{b}) = \frac{e_a}{R - \boldsymbol{R} \cdot \dot{\boldsymbol{a}}} \sum_{j=1,2} [\boldsymbol{\alpha}^{(j)} \cdot \dot{\boldsymbol{a}}]\boldsymbol{\alpha}^{(j)}. \tag{44}$$

We Fourier-analyze (44) to get

$$\boldsymbol{A}_{\text{ret}}^{(a)}(\boldsymbol{b}) = \sum_{l=-\infty}^{\infty} \sum_{j=1,2} [\boldsymbol{\alpha}^{(j)} \cdot \boldsymbol{A}_l]\boldsymbol{\alpha}^j \exp\left(-\frac{2\pi i l t}{T'}\right), \tag{45}$$

where

$$A_l = \frac{e_a}{RT'} \int \frac{\dot{\boldsymbol{a}}}{1 - \dot{\boldsymbol{a}} \cdot \boldsymbol{R}/R} \exp\left(\frac{2\pi i l t'}{T}\right) dt'. \tag{46}$$

The range T' of t' is given by

$$t' = t + R + \frac{1}{R} \boldsymbol{R} \cdot (\boldsymbol{b} - \boldsymbol{a}), \qquad 0 \le t \le T. \tag{47}$$

Since $\boldsymbol{b}$, $\boldsymbol{R}$, are not to be regarded as varying with t, we have

$$\frac{dt'}{dt} = 1 - \frac{\dot{\boldsymbol{a}} \cdot \boldsymbol{R}}{R}, \tag{48}$$

so that changing the variable from t' to t in (46) gives

$$A_l = \frac{e_a}{RT'} \exp\left[\frac{2\pi i l}{T'}\left(R + \frac{\boldsymbol{R} \cdot \boldsymbol{b}}{R}\right)\right] \int_0^T \dot{\boldsymbol{a}} \exp\left[\frac{2\pi i l}{T'}\left(t - \frac{\boldsymbol{a} \cdot \boldsymbol{R}}{R}\right)\right] dt, \tag{49}$$

with

$$T' = T - \frac{\boldsymbol{R}}{R} [\boldsymbol{a}(T) - \boldsymbol{a}(0)]. \tag{50}$$

The effect of dispersion in the cosmological medium is to introduce both a phase change and damping into (49), modifying it to

$$A_l = \frac{e_a}{RT'} \exp\left\{-\frac{1}{2}\tau_l + i\left[\frac{2\pi l}{T'}\left(R + \frac{\boldsymbol{R} \cdot \boldsymbol{b}}{R}\right) + \chi_l\right]\right\}$$
$$\int_0^T \dot{\boldsymbol{a}} \exp\left[\frac{2\pi i l}{T'}\left(t - \frac{\boldsymbol{a} \cdot \boldsymbol{R}}{R}\right)\right] dt, \tag{51}$$

the phase change being expressed by χ_l and the damping by τ_l.

We now substitute (45) into (40). We shall make use of the fact (to be shown later) that χ_l is a very large phase angle, and use this to wipe out any cross products of $\boldsymbol{A}_l$, $\boldsymbol{A}_{l'}$ which arise in (39), except those with $l' = -l$, giving

$$\frac{e_a^2 e_b^2}{3R^2 T'^2} \frac{(E_f - E_i)^2}{\hbar^4} \sum_{l=-\infty}^{\infty} \left| \boldsymbol{b}_{if}\left(\frac{2\pi l}{T'}\right)\right|^2 e^{-\tau_l} \left| \int_0^{T'} \exp\left[\frac{it'}{\hbar}\left(E_f - E_i - \frac{2\pi l \hbar}{T'}\right)\right] dt' \right|^2$$
$$\times \sum_{j=1,2} \int_0^T \boldsymbol{\alpha}^{(j)} \cdot \dot{\boldsymbol{a}} \exp\left[\frac{2\pi l i}{T'}\left(t - \frac{\boldsymbol{a} \cdot \boldsymbol{R}}{R}\right)\right] dt$$
$$\times \int_0^T \boldsymbol{\alpha}^{(j)} \cdot \dot{\boldsymbol{a}}' \exp\left[\frac{2\pi i l}{T'}\left(\frac{\boldsymbol{a}' \cdot \boldsymbol{R}}{R} - t\right)\right] dt, \tag{52}$$

in which we have already averaged with respect to all orientations of b to remove product terms in the components of $\boldsymbol{\alpha}^{(1)}$ and $\boldsymbol{\alpha}^{(2)}$, and where $\boldsymbol{b}_{if}\left(\frac{2\pi l}{T'}\right)$ is the matrix element of $\boldsymbol{b}$ $\exp\left[i\frac{2\pi l}{T'}\frac{\boldsymbol{R}\cdot\boldsymbol{b}}{R}\right]$ with respect to ψ_i and ψ_f. We have also ignored variations of $\boldsymbol{b}$ over Δ_+ in comparison with T'. Expression (52) can be simplified further by the results

$$\left|\int_0^{T'} \exp\left[\frac{it'}{\hbar}\left(E_f - E_i - \frac{2\pi\hbar l}{T'}\right)\right] dt'\right|^2 \simeq 2\pi T'\hbar\, \delta\left(E_f - E_i - \frac{2\pi l\hbar}{T'}\right) \tag{53}$$

$$\sum_{l=-\infty}^{\infty} \frac{\sin^2\alpha}{(\pi l - \alpha)^2} = 1. \tag{54}$$

Writing

$$\hbar k = E_f - E_i,\ \boldsymbol{k} = \frac{k\boldsymbol{R}}{R}, \tag{55}$$

and using (53) and (54), we reduce (52) to the form

$$\frac{e_a{}^2 e_b{}^2 k^2}{3R^2\hbar^2} e^{-\tau(k)} |\boldsymbol{b}_{if}(k)|^2 \sum_{j=1,2} \int_0^T \boldsymbol{\alpha}^{(j)} \cdot \dot{\boldsymbol{a}} e^{-i\boldsymbol{k}\cdot\boldsymbol{a} + ikt}\, dt$$

$$\times \int_0^T \boldsymbol{\alpha}^{(j)} \cdot \dot{\boldsymbol{a}}' e^{i\boldsymbol{k}\cdot\boldsymbol{a}' - ikt}\, dt = X \text{ (say)}. \tag{56}$$

Suppose that at a distance R in a particular solid angle $d\Omega$, there are $n(k)\,dk$ particles per unit volume with states f and i such that $(E_f - E_i)/\hbar$ lies in the range k to $k + dk$. Writing $\overline{|\boldsymbol{b}_{if}(k)|^2}$ for the average of $|\boldsymbol{b}_{if}(k)|^2$ for all systems satisfying this requirement, the contribution to $F[\boldsymbol{a}, \boldsymbol{a}']$ from all absorbers between R and $R + dR$ and in $d\Omega$ is

$$[1 + X]^{n(k)\,dk\,R^2\,dR\,d\Omega}. \tag{57}$$

Since X is very small and the index is large, we can rewrite (57) in the form

$$\exp[X \cdot n(k)\,dk\,R^2\,dR\,d\Omega]. \tag{58}$$

The function $\tau(k)$ in (56) is just the optical depth of the absorbing medium at frequency k. It is not hard to show that

$$\frac{d\tau(k)}{dR} = \frac{4\pi^2}{3}\frac{e_b{}^2 k}{\hbar}\overline{|\boldsymbol{b}_{if}(k)|^2} n(k), \tag{59}$$

the damping expressed by $\tau(k)$ being due to induced upward transitions in the absorber. Thus (59) follows from first-order perturbation theory. Remembering that absorber particles contribute as a product, as in (34), we next integrate the exponent of (58) with respect to R and with respect to k. Using (59), and letting $\tau \to \infty$ as $R \to \infty$, we obtain

$$\exp\left[\frac{e_a^{\,2}}{4\pi^2\hbar}\, d\Omega \int_0^\infty k\, dk \sum_{j=1,2} \int_0^T (\boldsymbol{\alpha}^{(j)} \cdot \dot{\boldsymbol{a}}) e^{-i\boldsymbol{k}\cdot\boldsymbol{a}+ikt}\, dt \cdot \int_0^T (\boldsymbol{\alpha}^{(j)} \cdot \dot{\boldsymbol{a}}') e^{i\,\boldsymbol{k}\cdot\boldsymbol{a}'-ikt}\, dt\right]. \tag{60}$$

It can be verified, again without difficulty, that the part (60) of the influence functional leads to the usual spontaneous transition formula. Also, since the integration is over positive values of k only, we do not have spontaneous upward transitions.

To complete the present problem, we still have to consider the terms in (38) involving $[\int_{\Delta_+}]^2$ and $[\int_{\Delta_+'}]^2$. After an analysis similar to that given above, we finally get

$$\begin{aligned} F[\boldsymbol{a}, \boldsymbol{a}'] = \exp\Bigg\{\frac{e_a^{\,2}}{4\pi^2\hbar} \int d\Omega \int_0^\infty k\, dk \\ \cdot \sum_{j=1,2} \Bigg[\int_0^T (\boldsymbol{\alpha}_k^{(j)} \cdot \dot{\boldsymbol{a}}) e^{-i\boldsymbol{k}\cdot\boldsymbol{a}+ikt}\, dt \int_0^T (\boldsymbol{\alpha}_k^{(j)} \cdot \dot{\boldsymbol{a}}') e^{i\boldsymbol{k}\cdot\boldsymbol{a}'-ikt}\, dt \\ - \int_0^T (\boldsymbol{\alpha}_k^{(j)} \cdot \dot{\boldsymbol{a}}) e^{-i\boldsymbol{k}\cdot\boldsymbol{a}-ikt}\, dt \int_0^t (\boldsymbol{\alpha}_k^{(j)} \cdot \dot{\boldsymbol{a}}) e^{i\boldsymbol{k}\cdot\boldsymbol{a}+ik\tilde{t}}\, d\tilde{t} \\ - \int_0^T (\boldsymbol{\alpha}_k^{(j)} \cdot \dot{\boldsymbol{a}}') e^{-i\boldsymbol{k}\cdot\boldsymbol{a}'+ikt}\, dt \int_0^t (\boldsymbol{\alpha}_k^{(j)} \cdot \dot{\boldsymbol{a}}') e^{i\boldsymbol{k}\cdot\boldsymbol{a}'-ik\tilde{t}}\, d\tilde{t}\Bigg]\Bigg\}. \end{aligned} \tag{61}$$

Here we have also integrated with respect to solid angle and have added the subscript $\boldsymbol{k}$ to $\boldsymbol{\alpha}^{(j)}$ to remind us that $\boldsymbol{\alpha}^{(j)}$ is a unit vector perpendicular to $\boldsymbol{k}$.

It is possible to arrive at exactly the same influence functional by a path-integral quantization of Maxwell fields. The present theory therefore is equivalent to the Maxwell theory in its quantum description of electromagnetic phenomena that arise from nonrelativistic motions of charged particles. Just as in the classical case, the parameters that characterize the absorbing medium do not appear in the final result. The terms in (61) additional to (60) describe level shifts.

§4.4. THE ROLE OF COSMOLOGY (LECTURE 9)

The expansion of the universe was not explicitly mentioned in the above analysis, although we used the important cosmological property of complete absorption in the future, by letting $\tau \to \infty$ as $R \to \infty$. We now point out some of the cosmological issues involved in the calculation. To fix ideas, we choose the steady-state model as an example, since it satisfies the necessary classical requirement of perfect absorption in the future and of imperfect absorption in the past. We shall use the conformal transformations described in the classical discussion, except that the time coordinates t and τ will be interchanged to keep the same notation as in §4.3. Thus the Robertson-Walker line element will be written as

$$ds^2 = d\tau^2 - e^{2H\tau}[dr^2 + r^2(d\theta^2 + \sin^2\theta \, d\phi^2)], \tag{62}$$

and its conformally flat form as

$$ds^2 = \left(-\frac{1}{Ht}\right)^2 [dt^2 - dr^2 - r^2(d\theta^2 + \sin^2\theta \, d\phi^2)]. \tag{63}$$

Because of the conformal invariance of electromagnetism, we used (r, θ, ϕ, t) coordinates in §4.3, just as if we were working in flat space.

In the spontaneous transition $E_m \to E_n$ of the local system, the important frequencies involved in the influence functional (61) are those in the neighborhood of

$$k = (E_m - E_n)/\hbar. \tag{64}$$

This frequency has to be matched by the value of $(E_f - E_i)/\hbar$ for the absorber transitions $\psi_i \to \psi_f$. If, however, we wish to consider E_i, E_f, with respect to proper time at the absorber, it is necessary to take account of the redshift effect of the expansion of the universe. Thus when a wave, starting with frequency k at the source, reaches an absorber at coordinate r, the proper frequency has become

$$\omega = (1 - Hr)k, \tag{65}$$

and E_i, E_f, with respect to local proper time are related to k by

$$E_f - L_i = \hbar k(1 - Hr). \tag{66}$$

If we displace the absorber by a proper distance dl away from the source, the frequency in tune with the wave changes by $d\omega$, where

$$d\omega = -Hk\,dr, \qquad dr = (1 - Hr)\,dl. \tag{67}$$

The second relation in (67) follows from the conformal transformation. Hence we get

$$\frac{d\omega}{\omega} = -H\,dl. \tag{68}$$

Suppose the absorbers are effective over a range $[\omega_{min}, \omega_{max}]$ of frequencies. Then if $k > \omega_{max}$, the range of l over which the absorbers can be in tune with k is

$$l = \int dl = H^{-1} \int_{\omega_{min}}^{\omega_{max}} \frac{d\omega}{\omega} = H^{-1}\, ln \frac{\omega_{max}}{\omega_{min}}. \tag{69}$$

For complete absorption, we need $l \to \infty$. Hence we need a process with $\omega_{min} \to 0$. Collisional absorption provides such a process.

The classical attenuation for a wave of frequency ω over a length dl caused by collisions with frequency ν is

$$\exp\left[-\frac{2\pi\nu Ne^2}{m\omega^2}\,dl\right], \tag{70}$$

where N is the number density of electrons, e the electronic charge, and m the electronic mass. Using (65) and (67), the damping produced over the coordinate range $0 \le r \le R$ is

$$\exp\left[-\frac{2\pi\nu Ne^2}{mk^2}\int_0^R \frac{dr}{(1 - Hr)^3}\right]. \tag{71}$$

For large R, appreciable damping occurs when

$$(1 - Hr)^2 \simeq \frac{2\pi\nu Ne^2}{mk^2H}, \tag{72}$$

which corresponds to an effective proper frequency

$$\omega_{eff}^2 \simeq \frac{2\pi\nu Ne^2}{mH}. \tag{73}$$

For ionized hydrogen, ν at ω_{eff} is given by

$$\nu = 2\pi Nv\left(\frac{e^2}{mv^2}\right)^2 ln\left(\frac{mv^2}{\hbar\omega_{eff}}\right), \tag{74}$$

where v is a typical electron velocity. Taking $H^{-1} \sim 3 \cdot 10^{17}$ sec, $N \sim 10^{-5}$ cm^{-3}, $v \simeq \frac{1}{300}$ (of velocity of light), and substituting for e, m, k, in (73) and (74), we can solve for ω_{eff} and ν:

$$\omega_{\text{eff}} \simeq 10^6 \text{ sec}^{-1}, \qquad \nu \simeq 10^{-10} \text{ sec}^{-1}. \tag{75}$$

Thus a wave with frequency greater than 10^6 sec^{-1} is first redshifted to 10^6 sec^{-1} and then absorbed. The effective absorption takes place over a distance of the order of 10^{28} cm. A wave with frequency less than 10^6 sec^{-1} is absorbed without having to be redshifted.

From (75) we find that the dimensionless parameter

$$\frac{4\pi N e^2}{m\omega_{\text{eff}}^2} \simeq 10^{-7}. \tag{76}$$

Since the real part of refractive index is given by

$$n \simeq 1 - \frac{2\pi N e^2}{m\omega_{\text{eff}}^2}, \tag{77}$$

we see that $1 - n \simeq 10^{-7}$.

Suppose we take two Fourier components k and $k + \Delta k$ given by neighboring values of l in the analysis of §4.3,

$$k = \frac{2\pi l}{T'}, \qquad k + \Delta k = \frac{2\pi(l+1)}{T'}. \tag{78}$$

These are in tune with absorber frequencies ω, $\omega + \Delta\omega$, where

$$\frac{\Delta\omega}{\omega} = \frac{\Delta k}{k} = \frac{1}{l}. \tag{79}$$

The difference in n for these two neighboring frequencies is

$$\Delta n = \frac{4\pi N e^2}{m\omega^3}\,\Delta\omega \simeq \frac{10^{-7}}{l} \tag{80}$$

when $\omega \simeq \omega_{\text{eff}}$. Over a distance H^{-1} this causes a phase difference

$$\Delta n \omega_{\text{eff}} H^{-1} \simeq 10^{-7}\, \omega_{\text{eff}} H^{-1} \left(\frac{2\pi}{kT'}\right). \tag{81}$$

For an allowed transition, $T'k$ does not exceed unity by a factor of more than $(e^2/\hbar)^{-3}$. Thus the phase difference is greater than

$\sim 10^{-12}\,\omega_{\text{eff}} H^{-1} \simeq 10^{11}$ radians. This explains why even neighboring frequencies become randomly phased, and why we were justified in omitting crossproducts in (39). Thus random phasing is of cosmological origin, and does not have to be assumed *ad hoc* as in the field theory.

These numbers could change somewhat if other cosmological models with appropriate absorption properties were used, but the basic ideas would remain the same.

§4.5. THE PHOTON AND THE THERMODYNAMIC ARROW OF TIME

How does the photon fit into the above picture? First, we note that when an atomic electron jumps down, it loses energy $E_m - E_n = \hbar k$. The energy is not emitted classically in an isotropic way, but as the influence functional (61) shows, it is emitted in a certain direction in a solid angle $d\Omega$ with a certain probability. We may describe this exchange of energy between the source and an absorber in $d\Omega$ as an exchange of a photon of energy $\hbar k$.

Suppose further that there is an external direct-particle field present, as given by the first term on the righthand side of (31). The electron then jumps up and down between the states with energies E_n, E_m with probabilities which satisfy

$$P(E_m \rightarrow E_n) = P(E_n \rightarrow E_m) + \text{probability of spontaneous transition } E_m \rightarrow E_n.$$

Now we can write

$$\frac{P(E_m \rightarrow E_n)}{P(E_n \rightarrow E_m)} = \frac{n+1}{n} \tag{82}$$

where n is some number. Since the induced transition probabilities are proportional to the intensity of the incident field, we can say that the intensity is that of n photons.

In this definition of the "number of photons" in the external field, we use the spontaneous transition probability as an absolute unit. Indeed, this definition of the photon is physically more realistic than the abstract one given by the quantization rules of field theory. Physically, a photon is seen either when it is emitted or absorbed—i.e., when it begins or ends on a particle world line, and the process of measurement of its energy causes transitions in the measuring equipment. It is this effect that we are using here to define the photon.

Equation (82) also gives us a clue to the connection between the arrows of time in cosmology and thermodynamics. As is well known, (82) leads to the blackbody spectrum of thermodynamics; yet (82) arises here from a cosmological asymmetry which permits only downward spontaneous transitions. If the cosmological medium could act as a maser, augmenting the outgoing wave systematically instead of absorbing it, we could achieve spontaneous upward transitions. It may be possible to formulate a consistent theory with spontaneous upward transitions in a contracting universe which gets hotter in the future. In that case one would expect "population inversion" and thermodynamics going "backwards." Further work is necessary to investigate this possibility. If successful we could argue that there is a strong interrelation of the time arrows of cosmology, thermodynamics, and electrodynamics.

This completes the nonrelativistic discussion. We now turn to the interactions of relativistic spin-half particles. We shall put $\hbar = 1$ from now on, thereby eliminating a redundant unit.

§4.6. PATH AMPLITUDES FOR DIRAC PARTICLES (LECTURE 10)

There are two ways of approaching the problem of defining path amplitudes for relativistic spin-half particles. The first one, making use of the Feynman method based on (4), presents difficulties at first, since we have no clear way of writing S for such particles. The formula

$$S = -\int m_a \, da \tag{83}$$

is not suitable in itself, since it does not contain spin. We shall consider this problem later.

The second approach consists of extending (8), using a suitable propagator. Direct analogy with the nonrelativistic case suggests that we take $K(2; 1)$ as the solution of the inhomogeneous Dirac equation

$$\left[\gamma^i \frac{\partial}{\partial x_2{}^i} + im\right] K(2; 1) = \delta_4(2, 1), \tag{84}$$

which satisfies the boundary condition

$$K(2; 1) = 0 \qquad \text{for } t_2 < t_1. \tag{85}$$

Using the free-particle solutions u_n of the Dirac equation, we can write K explicitly as

$$K_0^+(2; 1) = \theta(t_2 - t_1) \sum_n u_n(2)\bar{u}_n(1), \tag{86}$$

where θ is the heaviside function and $\bar{u}_n = u_n^* \gamma_4$. We shall also use the notation

$$\gamma^i A_i = \not{A}. \tag{87}$$

The sum (86) is over all energy states, negative as well as positive. This poses the usual difficulty of particles with negative energy states. Feynman, who used $K_0^+(2; 1)$ in defining path amplitudes, had to postulate a vacuum filled with negative-energy particles. Thus even to describe the motion of a single particle using (86), we have to take into account a background with an infinity of negative-energy particles.

To avoid this situation, we proceed in a different way. We introduce two types of particles—those with positive energy always going forward in time, and those with negative energy always going backward in time. These may subsequently be identified with particles and antiparticles, e.g., electrons and positrons.

We therefore introduce two types of paths, Γ^+ and Γ^- (see Figure 4.3). The Γ^+ are always going forward in time, and amplitude along them is defined by (8), using K_0^+:

$$P(\Gamma^+) = \lim_{N\to\infty} \prod_{r=1}^{N} A_r K_0^+(Q_r; Q_{r-1}). \tag{88}$$

The Γ^- paths are always going backward in time, with amplitudes

$$P(\Gamma^-) = \lim_{N\to\infty} \prod_{r=1}^{N} A_r K_0^-(Q_r; Q_{r-1}), \tag{89}$$

where the points Q_r form a reversed chronological sequence, and K_0^- is given by

$$K_0^-(2; 1) = -\theta(t_1 - t_2) \sum_n u_n(2)\bar{u}_n(1). \tag{90}$$

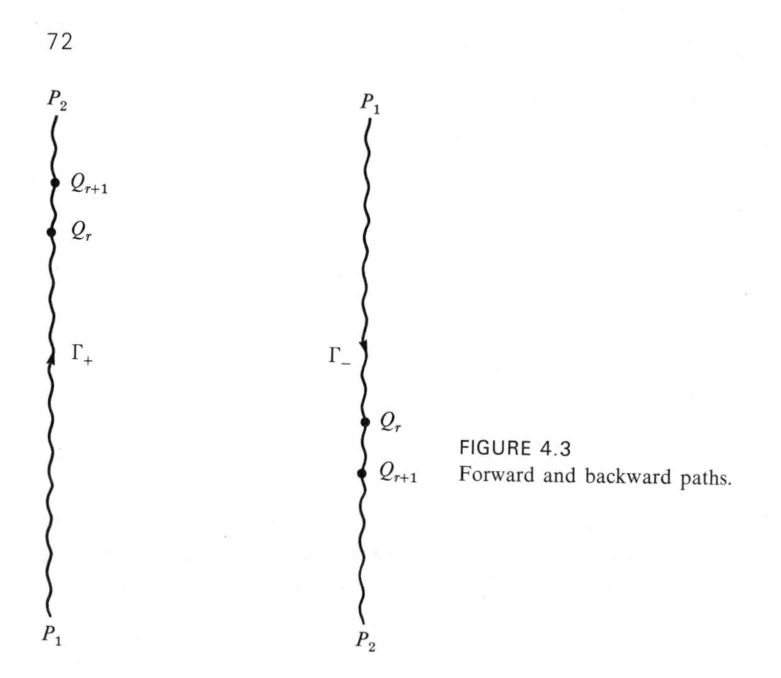

FIGURE 4.3
Forward and backward paths.

We make a similar distinction for wave functions. Suppose $\psi(\boldsymbol{x}, t)$ is given on any spacelike surface. We then write

$$\psi(\boldsymbol{x}, t) = \psi_+(\boldsymbol{x}, t) + \psi_-(\boldsymbol{x}, t), \tag{91}$$

where ψ_+ is made up of positive energy states only and ψ_- is made up of negative energy states. The ψ_+ part is propagated forward by the Γ^+ paths while the ψ_- part is propagated backward by Γ^- paths.

Suppose we have a slab of spacetime defined by $t_1 \leq t \leq t_2$. Let ψ_+ be given on $t = t_1$ and ψ_- on $t = t_2$. Then using arguments similar to those in the nonrelativistic path-integral theory, we get $\psi(\boldsymbol{x}, t)$ as

$$\psi(\boldsymbol{x}, t) = \int K_0{}^+(\boldsymbol{x}, t; \boldsymbol{x}_1, t_1)\gamma_4\psi_+(\boldsymbol{x}_1, t_1)\, d^3\boldsymbol{x}_1 - \int K_0{}^-(\boldsymbol{x}, t; \boldsymbol{x}_2, t_2)\gamma_4\psi_-(\boldsymbol{x}_2, t_2)\, d^3\boldsymbol{x}_2. \tag{92}$$

The γ_4 factor arises from the necessity to preserve invariance. Equation (92) can be rewritten in the form

$$\psi(\boldsymbol{x}, t) = \int K_+(\boldsymbol{x}, t; \boldsymbol{x}_1, t_1)\gamma_4\psi(\boldsymbol{x}_1, t_1)\, d^3\boldsymbol{x}_1 - \int K_+(\boldsymbol{x}, t; \boldsymbol{x}_2, t_2)\gamma_4\psi(\boldsymbol{x}_2, t_2)\, d^2\boldsymbol{x}_2, \tag{93}$$

where we no longer restrict ψ to the positive energy component on $t = t_1$ and to the negative energy component on $t = t_2$, but define a new propagator

$$K_+(2, 1) = \begin{cases} \sum_{E_n > 0} u_n(2)\bar{u}_n(1), & t_2 > t_1, \\ \\ -\sum_{E_n < 0} u_n(2)\bar{u}_n(1), & t_2 < t_1. \end{cases} \tag{94}$$

Thus in the first integral on the righthand side of (93), only the positive-energy part ψ_+ is propagated to $(\boldsymbol{x}, t)$. Similarly, only the negative-energy part is propagated to $(\boldsymbol{x}, t)$ from $(\boldsymbol{x}_2, t_2)$.

To summarize, the total amplitude at a point within the slab $t_1 \leq t \leq t_2$ is determined by Γ^- paths from the future as well as by Γ^+ paths from the past. Since $K_0^{\pm}$ are zero outside the light cones, $\Gamma^{\pm}$ paths are always timelike. We can build up more complicated paths (see Figure 4.4) by joining Γ^+ and Γ^- paths in any order. For propagation of a free particle, however, such zigzag paths make no contribution, as can be easily verified from the above definitions.

Next we introduce the electromagnetic potential B_i. The path amplitude for a charged particle a is now given for a Γ^+ path by

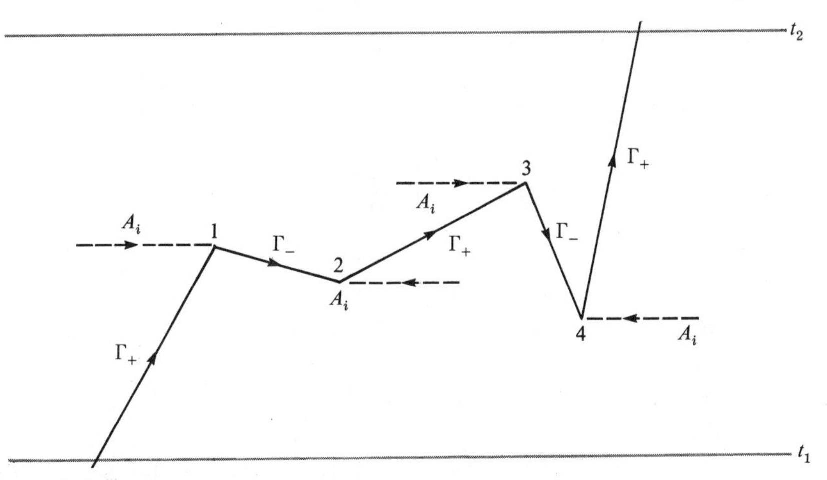

FIGURE 4.4
Path reversals caused by the electromagnetic field A_i. If A_i were zero at a reversal point, the path would make no contribution to the amplitude.

$$P^B(\Gamma^+{}_{21}) = P(\Gamma^+{}_{21}) \exp \int_{\Gamma^+{}_{21}} -ieB_i \, da^i, \tag{95}$$

where $P(\Gamma^+{}_{21})$ is the free-particle amplitude along $\Gamma^+{}_{21}$ from 1 to 2. $P^B(\Gamma^-{}_{21})$ is similarly defined.

If we consider the same problem as above, i.e., given ψ on two planes $t = t_1$ and $t = t_2 > t_1$, to calculate $\psi(\boldsymbol{x}, t)$ at an interior point of the slab $t_1 \leq t \leq t_2$, we get, instead of (93),

$$\psi(\boldsymbol{x}, t) = \int K^B{}_+(\boldsymbol{x}, t; \boldsymbol{x}_1, t_1)\gamma_4\psi(\boldsymbol{x}_1, t_1)\, d^3\boldsymbol{x}_1 - \int K^B{}_+(\boldsymbol{x}, t; \boldsymbol{x}_2, t_2)\gamma_4\psi(\boldsymbol{x}_2, t_2)\, d^3\boldsymbol{x}_2, \tag{96}$$

where

$$K^B{}_+(2; 1) = K_+(2; 1) - ie \int K_+(2; 3)\not{B}(3)K_+(3; 1)\, d\tau_3 + (-ie)^2 \int\int K_+(2; 3)\not{B}(3)K_+(3; 4)\not{B}(4)K_+(4; 1)\, d\tau_3\, d\tau_4 + \ldots. \tag{97}$$

(This result is derived in Appendix 3.) $K^B{}_+(2, 1)$ satisfies the equation

$$(\not{\nabla}_2 + ie\not{B}(2) + im)K^B{}_+(2; 1) = \delta_4(2, 1). \tag{98}$$

Many-Particle Systems

For a single particle we write

$$K^B{}_+(2, 1) = \int P(\Gamma_{21}) \exp\left(\int - ieB_i\, da^i\right)\mathcal{D}\Gamma_{21}, \tag{99}$$

where Γ_{21} can be made up out of any number of Γ^+ and Γ^- paths joined in any order. We shall consider (99) further later.

For a system of many particles we have a multiple-path integral. To write it, we change our notation slightly, denoting the initial and final position of the ath particle by a_i and a_f, respectively. Chronologically, a_i may not occur before a_f, since paths can go backward and forward. A typical path joining a_i to a_f will be denoted by Γ_a. Then the propagator for a non-interacting system of particles in a given field B_i is formed as a product of (99) taken for each particle,

$K^B{}_+[\ldots; a_f, a_i; \ldots] =$

$$\int\cdots\int \left[\prod_a P(\Gamma_a)\right] \exp\left(\sum_a -ie_a \int B_i\, da^i\right) \prod_a \mathscr{D}\Gamma_a. \quad (100)$$

If, on the other hand, the particles are mutually interacting, (100) is changed to

$K^{(B+R)}{}_+[\ldots; a_f, a_i; \ldots] =$

$$\int\cdots\int \prod_a P(\Gamma_a) \exp\left(\sum_a -ie_a \int B_i\, da^i\right) \cdot \exp(iR) \prod_a \mathscr{D}\Gamma_a, \quad (101)$$

where

$$R = -\frac{1}{2}\sum_a \sum_b e_a e_b \int_{\Gamma_a} \int_{\Gamma_b} \delta(s^2{}_{AB})\, da_i\, db^i. \quad (102)$$

In (102) we have apparently included-self action. We shall return to this point in some detail later.

We note that if we are dealing with similar particles, the exclusion principle requires the initial wavefunction to be antisymmetric. If the initial wavefunction is $\Psi[\ldots, a_i, \ldots]$, then the final wavefunction given by

$\Psi[\ldots, a_f, \ldots] =$

$$\int\cdots\int K^{(B+R)}{}_+[\ldots; a_f, a_i; \ldots] \prod_a \not{n}_a \cdot \Psi[\ldots, a_i, \ldots] \prod_a dS_a \quad (103)$$

is also antisymmetric. In (103), each a_i runs over a closed three-dimensional surface, with normal $n_a{}^i$ and surface element dS_a at a typical point.

Equation (103) requires us to antisymmetrize the amplitude at each stage. Thus (see Figure 4.5) if two electrons start at 1, 2, and end at 3, 4, we have to calculate amplitudes along paths Γ_{31}, Γ_{42}, multiply them, and then subtract the amplitude product for paths Γ_{41}, Γ_{32}. This immediately leads to closed-loop paths in perturbation expansions. For example (see Figure 4.5), if the electron goes from 1 to 3, is scattered back to 2 and then forward

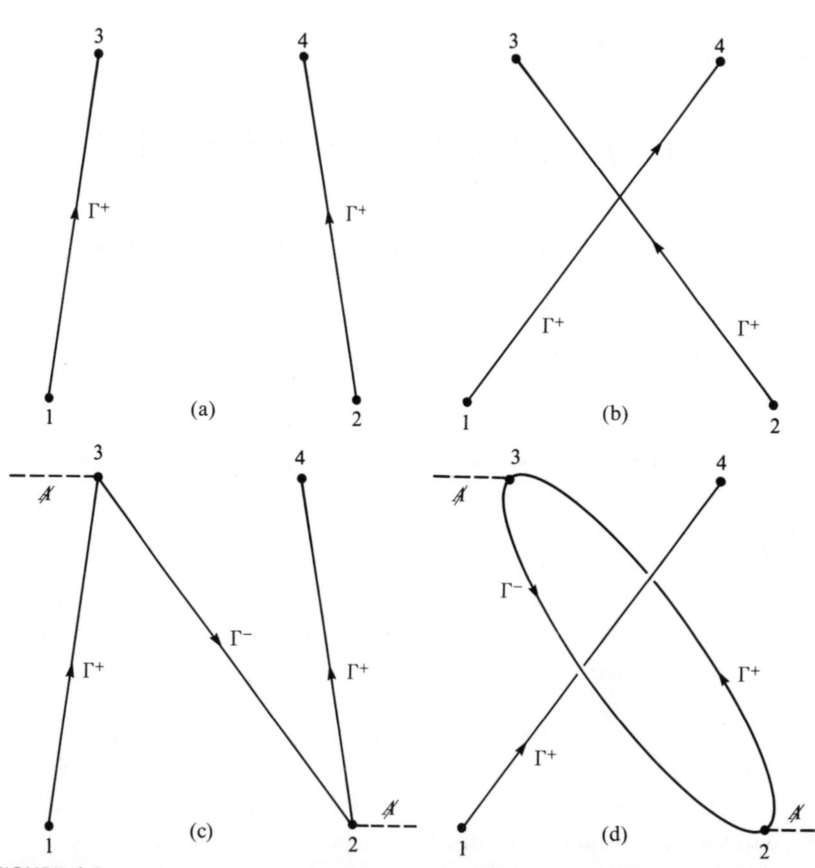

FIGURE 4.5
In two-particle scattering, the exclusion principle requires the amplitude for (b) to be subtracted from that of (a). The same principle, when applied to multiple scattering of a "single" particle by electromagnetic potentials, leads to closed loops, as shown in (c) and (d).

to 4, we must subtract from the amplitude along the path

$$\Gamma_{42}\Gamma_{23}\Gamma_{31} \tag{104}$$

the amplitude along

$$\Gamma_{41}\Gamma_{23}\Gamma_{32}. \tag{105}$$

In (105) we have a particle moving freely from 1 to 4 and a closed loop at which the potential acts twice.

A closed loop, Γ^0, is made up of $\Gamma^{\pm}$ paths, and its amplitude can be similarly defined. However, since here the ends are identified, the amplitude becomes a trace:

$$P^B(\Gamma^0) = \mathrm{Tr}\left[\prod_i P^B(\Gamma_{i;i-1})\right]. \tag{106}$$

Thus P^B can be evaluated starting from any point of Γ^0. Equation (106) also gives the relation

$$P^B(\Gamma^0) = \exp\left(-ie\oint_{\Gamma^0} B_i\, dl^i\right) P(\Gamma^0), \tag{107}$$

where dl^i is an element of length along Γ^0 and e is the charge of the particle giving rise to the loop Γ^0. We shall use these results in the discussion of vacuum polarization.

We come now to the status of the self-action term in (102). Once we allow paths to go backward or forward, we cannot properly identify two distant sections of world lines as belonging to distinct particles. This inability requires the $\delta(s^2)$ interaction to act between *all* sections. In the nonrelativistic theory, all paths are forward-directed and are timelike. The interaction $\delta(s^2{}_{AB})$ shown in Figure 4.6 therefore vanishes unless A and B coincide. So long as we avoid this possibility, classical theory and nonrelativistic quantum theory remain unchanged by the inclusion of self-action in (102).

We therefore make the rule that in the self-action we do not permit the two "legs" of the delta function to coincide. We may allow them to approach to a certain minimum distance. If we go to smaller separations, we would bring in true self-action, i.e., a particle interacting with itself. We shall give a quantitative expression to this idea in the next section.

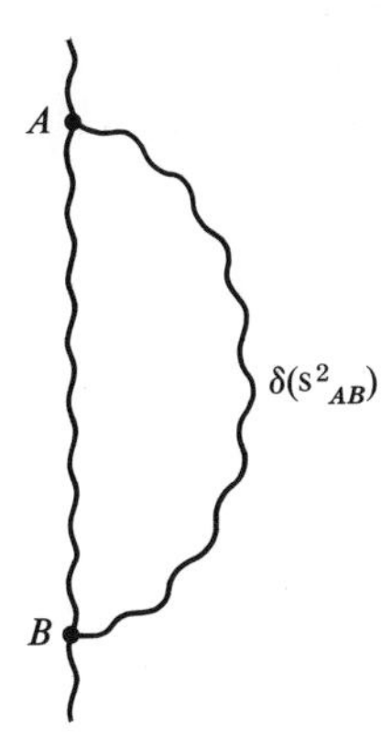

FIGURE 4.6
The self-action described by the propagator $\delta(s^2{}_{AB})$ is zero when the separation of points A, B, is timelike.

§4.7. THE QUANTUM RESPONSE OF THE UNIVERSE (LECTURE 11)

In the multiple-path integral (101), the electromagnetic phase factor is i times

$$-\sum_a e_a \int B_i \, da^i - \frac{1}{2} \sum_a \sum_b e_a e_b \int_{\Gamma_a} \int_{\Gamma_b} \delta(s^2{}_{AB}) \, da_i \, db^i. \qquad (108)$$

For definiteness, we imagine a slab $0 \leq t \leq T$, taking the paths Γ_a, Γ_b, to be segments inside the slab. There are interactions with segments outside the slab as well, and these determine B_i. Thus we can write

$$B^i = B^i{}_{t<0} + B^i{}_{t>T}, \qquad (109)$$

where the two terms on the righthand side correspond to effects of the past ($t < 0$) and of the future ($t > T$), respectively. We have seen that $B^i{}_{t<0}$ does not have any complicated properties—it represents the retarded fields of other particles in the universe. It is the future component $B^i{}_{t>T}$ that has the interesting properties. So to simplify discussion we will take

$$B^i{}_{t<0} = 0 \qquad (110)$$

and concentrate attention on $B^i{}_{t>T}$.

We begin with some definitions. We have already defined the direct-particle potential due to particle b:

$$A_i^{(b)}(X) = \int e_b \, \delta(s^2{}_{XB}) \, db_i. \qquad (111)$$

Separating $\delta(s^2{}_{XB})$ into advanced and retarded parts,

$$\delta(s^2{}_{XB}) = \frac{1}{2|\boldsymbol{x} - \boldsymbol{x}_B|} [\delta(t - t_B + |\boldsymbol{x} - \boldsymbol{x}_B|) + \delta(t - t_B - |\boldsymbol{x} - \boldsymbol{x}_B|)], \qquad (112)$$

where $X^i \equiv (\boldsymbol{x}, t)$, $B^i \equiv (\boldsymbol{x}_B, t_B)$, we define correspondingly

$$A_i^{(b)}(X) = \frac{1}{2} [A_i^{(b)}(X)_{\text{ret}} + A_i^{(b)}(X)_{\text{adv}}]. \qquad (113)$$

Let

$$A^i{}_{(a)}(X) = \sum_{b \neq a} A^{i(b)}(X) = \frac{1}{2} [A^i{}_{(a)}(X)_{\text{ret}} + A^i{}_{(a)}(X)_{\text{adv}}]. \qquad (114)$$

Thus

$$A^i{}_{(a)}(X)_{\text{ret}} = \sum_{b \neq a} e_b \int \frac{\delta(t - t_B - |\boldsymbol{x} - \boldsymbol{x}_B|)}{|\boldsymbol{x} - \boldsymbol{x}_B|} \, db^i,$$

$$A^i{}_{(a)}(X)_{\text{adv}} = \sum_{b \neq a} e_b \int \frac{\delta(t - t_B + |\boldsymbol{x} - \boldsymbol{x}_B|)}{|\boldsymbol{x} - \boldsymbol{x}_B|} \, db^i. \tag{115}$$

In the classical case, with the omission of the self-action terms, (108) can be written in the form

$$-\sum_a e_a \int B^i{}_{t>T} \, da_i - \frac{1}{2} \sum_a e_a \int [A^i{}_{(a)\text{ret}} + A^i{}_{(a)\text{adv}}] \, da_i. \tag{116}$$

We found in the classical case that, in a universe with consistent retarded solutions, the first term of (116) cancels the advanced component of the second term, doubles the retarded component, and produces the Dirac radiative reaction. Thus in the classical case the response of a completely absorbing universe may be expressed in the form

$$B_i(X)_{t>T} = \frac{1}{2} \sum_b [A_i^{(b)}(X)_{\text{ret}} - A_i^{(b)}(X)_{\text{adv}}]. \tag{117}$$

When we substitute (117) into (116), we get

$$-\sum_a e_a \int \left\{ A^i{}_{(a)\,\text{ret}} + \frac{1}{2} [A^{i(a)}{}_{\text{ret}} - A^{i(a)}{}_{\text{adv}}] \right\} da_i. \tag{118}$$

We now generalize (117) to quantum electrodynamics. To do so we refer back to the detailed nonrelativistic discussion in §4.3. To determine transition probabilities, we had to take the square of the modulus of certain path integrals. This brought in conjugate paths as well as paths. The appearance of both paths and conjugate paths was inevitable, since the local transitions came through the response of the universe, which in turn came through transitions in the absorber. In working out these transitions, one must make a probability, not an amplitude, calculation. Thus the response of the universe acts via both paths and conjugate paths.

The same situation occurs here. We have to take the square of the modulus of the path integral in (101). Corresponding to an integral over Γ_a, we have an integral over Γ'_a, the conjugate path. A typical expression in the exponential involving Γ_a and Γ'_a is i times

$$-\sum_a e_a \int_{\Gamma_a} \left[\frac{1}{2}(A^i{}_{\text{ret}} + A^i{}_{\text{adv}}) + B^i{}_{t>T}\right] da_i$$

$$+\sum_a e_a \int_{\Gamma'_a} \left[\frac{1}{2}(A'^i{}_{\text{ret}} + A'^i{}_{\text{adv}}) + B'^i{}_{t>T}\right] da'_i, \tag{119}$$

in which

$$\begin{aligned} A^i(X)_{\text{ret}} &= A^{i(a)}(X)_{\text{ret}} + A^i{}_{(a)}(X)_{\text{ret}}, \\ A^i(X)_{\text{adv}} &= A^{i(a)}(X)_{\text{adv}} + A^i{}_{(a)}(X)_{\text{adv}}. \end{aligned} \tag{120}$$

The difference from the classical situation is, as pointed out earlier, the inclusion of the self terms, those involving $A^{i(a)}$ in (120). To define the quantum response, we need the separation of various expressions into positive and negative frequencies. We shall use the following notation:

$$A^i(X)_{\text{ret}\pm} = \sum_b e_b \int \frac{\delta_\pm(t - t_B - |\boldsymbol{x} - \boldsymbol{x}_B|)}{2|\boldsymbol{x} - \boldsymbol{x}_B|}\, db^i,$$

$$A^i(X)_{\text{adv}\pm} = \sum_b e_b \int \frac{\delta_\pm(t - t_B + |\boldsymbol{x} - \boldsymbol{x}_B|)}{2|\boldsymbol{x} - \boldsymbol{x}_B|}\, db^i, \tag{121}$$

where

$$\delta_\pm(x) = \frac{1}{\pi}\int_0^\infty e^{\mp i\omega x}\, d\omega = \delta(x) \mp \frac{i}{\pi}\frac{\mathcal{P}}{x}. \tag{122}$$

(122) is used extensively in quantum electrodynamics. Its properties are discussed in Appendix 1 (p. 209). Thus we have

$$A^i{}_{\text{ret}} = A^i{}_{\text{ret}+} + A^i{}_{\text{ret}-}, \qquad A^i{}_{\text{adv}} = A^i{}_{\text{adv}+} + A^i{}_{\text{adv}-}. \tag{123}$$

We now give the quantum-mechanical analogue of (117):

$$B_i(X)_{t>T} = \frac{1}{2}\sum_b \{[A_i{}^{(b)}(X)_{\text{ret}+} - A_i{}^{(b)}(X)_{\text{adv}+}]$$

$$+ [A_i{}'^{(b)}(X)_{\text{ret}-} - A_i{}'^{(b)}(X)_{\text{adv}-}]\}. \tag{124}$$

The expression (124) remains unaltered if we interchange paths and conjugate paths and positive and negative frequencies. Thus

$$B_i'(X)_{t>T} = B_i(X)_{t>T}. \tag{125}$$

Also, when paths and conjugate paths coalesce, (124) reduces to (117). This happens in the classical limit when there is a unique path for each particle. Finally, (124) transforms relativistically as a 4-vector, although individual parts of (124) such as $A^i_{\text{ret}+}$ do not transform relativistically. It can be verified that, apart from the classical combination (123), the expressions in the square brackets in (124) are the only combinations which transform as a vector.

If we now consider a pair of particles a and b, and use (124) as the response of the universe, the electromagnetic phase factor involving a and b in the double path integral (see Appendix 3, p. 217) can be expressed in the form:

$$\begin{aligned}\exp\Bigg\{ie_a e_b\Bigg[&\iint\limits_{t_{A'}>t_B}\delta_+(s^2{}_{A'B})\,da'_i\,db^i - \iint\limits_{t_B>t_{A'}}\delta_-(s^2{}_{BA'})\,da'_i\,db^i\\ &+\iint\limits_{t_{B'}>t_A}\delta_+(s^2{}_{B'A})\,da^i\,db'_i - \iint\limits_{t_A>t_{B'}}\delta_-(s^2{}_{AB'})\,da^i\,db'_i\\ &-\iint\delta_+(s^2{}_{AB})\,da^i\,db_i + \iint\delta_-(s^2{}_{A'B'})\,da'_i\,db'^i\Bigg]\Bigg\}.\end{aligned} \tag{126}$$

This is the influence functional $F[a, b; a', b']$ for the interaction of particles a and b. If we use (126) we can treat the system as a local one in the slab $0 \leq t \leq T$.

The delta functions in (126) can be expressed as three-dimensional Fourier integrals. Thus $F[a, b; a', b']$ can be written in the form

$$\begin{aligned}F&[a, b; a', b']\\ &= \exp\Bigg(\frac{e_a e_b}{4\pi^2}\int d\Omega\int K\,dK\Bigg\{\iint \exp[-iK|t_A - t_B| + i\boldsymbol{k}\cdot(\boldsymbol{x}_B - \boldsymbol{x}_A)]\,da_i\,db^i\\ &\qquad + \iint \exp[iK|t_{A'} - t_{B'}| + i\boldsymbol{k}\cdot(\boldsymbol{x}_{A'} - \boldsymbol{x}_{B'})]\,da'_i\,db'^i\\ &\qquad - \iint \exp[iK(t_A - t_{B'}) + i\boldsymbol{k}\cdot(\boldsymbol{x}_{B'} - \boldsymbol{x}_A)]\,da_i\,db'^i\\ &\qquad - \iint \exp[iK(t_B - t_{A'}) + i\boldsymbol{k}\cdot(\boldsymbol{x}_{A'} - \boldsymbol{x}_B)]\,da'_i\,db^i\Bigg\}\Bigg).\end{aligned} \tag{127}$$

The terms in (127) have a simple interpretation. For $t_A > t_B$, the positive terms in the curly bracket contribute to a downward transition of particle

b and an upward transition of particle a, and vice versa for $t_B > t_A$. The negative terms contribute to downward transitions of both a and b. Since the coefficient in front of all terms is the same, all these processes occur with equal probability.

The case $a = b$ needs separate discussion, although the details are not seriously different. The influence functional is now given by (see Appendix 3, p. 217)

$$F[a, a'] = \exp\left\{ie^2{}_a\left[\iint_{t_{A'} > t_A} \delta_+(s^2{}_{A'A})\, da'_i\, da^i - \iint_{t_A > t_{A'}} \delta_-(s^2{}_{AA'})\, da'{}_i\, da^i \right.\right.$$
$$\left.\left. - \frac{1}{2}\iint \delta_+(s^2{}_{A\tilde{A}})\, da^i\, d\tilde{a}_i + \frac{1}{2}\iint \delta_-(s^2{}_{A'\tilde{A}'})\, da'^i\, d\tilde{a}_i'\right]\right\}. \quad (128)$$

Corresponding to (127), we have

$$F[a, a'] = \exp\left(\frac{e_a^2}{4\pi^2}\int d\Omega \int_0^\infty K\, dk \left\{-\iint \exp[iK(t_A - t_{A'})\right.\right.$$
$$+ i\boldsymbol{k}\cdot(\boldsymbol{x}_{A'} - \boldsymbol{x}_A)]\, da^i\, da'_i + \iint_{t_A > t_{\tilde{A}}} \exp[iK(t_{\tilde{A}} - t_A) + i\boldsymbol{k}\cdot(\boldsymbol{x}_{\tilde{A}} - \boldsymbol{x}_A)]\, da^i\, d\tilde{a}_i$$
$$\left.\left. + \iint_{t_{A'} > t_{\tilde{A}'}} \exp[iK(t_{A'} - t_{\tilde{A}'}) + i\boldsymbol{k}\cdot(\boldsymbol{x}_{A'} - \boldsymbol{x}_{\tilde{A}'})]\, da'^i\, d\tilde{a}'_i\right\}\right). \quad (129)$$

This is the same result as the one obtained earlier using the nonrelativistic theory. The first term in the curly bracket gives spontaneous transitions, whereas the second and third terms contribute radiative correction effects. Provided we add similar expressions for interactions involving closed loops, all of quantum electrodynamics is contained in (126) and (128). As explained in the introductory chapter, we regard this result as a critical requirement for any new point of view in physics. Having reached this position, we can now extend our work outside electrodynamics. The above discussion has been kept as brief as possible, giving only essentials; further details are given in Appendix 3.

Radiative Corrections

We end this chapter with remarks on radiative corrections in the direct-particle theory. Once again, Appendix 3 (p. 217) contains a more extensive discussion of these matters.

The main point on which radiative corrections depend is the fine structure of paths. We have already seen that $\Gamma^{\pm}$ paths are made up of chains of K_0 propagators. Now $K_0{}^+(2; 1)$ is given by

$$K_0{}^+(2;1) = \theta(t_2 - t_1)(\not{\nabla}_2 - im)\left[\frac{1}{2\pi}\,\delta(s^2{}_{21}) - \frac{m}{4\pi s_{21}}\,\theta(s^2{}_{21})J_1(ms_{21})\right]. \quad (130)$$

As the separation between 1 and 2 decreases, the delta function in (130) dominates. Thus a Γ^+ path is made up of small null segments—if we look at it sufficiently closely. If in the self-action integral

$$-\frac{1}{2}\,e_a{}^2 \int\int \delta(s^2{}_{A\tilde{A}})\,da^i\,d\tilde{a}_i \quad (131)$$

the points A and $\tilde{A}$ lie on the same null segment, $s^2{}_{A\tilde{A}} = 0$ and $d\tilde{a}_i\,da^i$ is proportional to $s^2{}_{A\tilde{A}}$. Since $\lambda\,\delta(\lambda) = 0$, we get no contribution provided A and $\tilde{A}$ are distinct. If A, $\tilde{A}$, are on different segments of a Γ^+ path, $s^2{}_{A\tilde{A}} > 0$ and $\delta(s^2{}_{A\tilde{A}}) = 0$. Thus the only contribution to (131) from such a path comes when A, $\tilde{A}$, are made to coincide. So long as we prevent this possibility, we should be able to avoid divergences. The same situation holds for Γ^- paths. On the other hand, we expect contributions to (131) from interactions between Γ^+ paths and Γ^- paths. These interactions are needed in the theory, otherwise there would be difficulty in understanding the interaction of particles and antiparticles.

In calculating the integrals involving self-action, we use the rule

$$|x^i{}_A - x^i{}_{\tilde{A}}| \geq |\varepsilon^i|,\; i = 1, 2, 3, 4, \quad (132)$$

to keep A and $\tilde{A}$ apart, where ε^i is a 4-vector associated with the particle. Explicitly, we take

$$\varepsilon^i = \varepsilon\,\frac{p^i}{m}, \quad (133)$$

where p^i is the 4-momentum, m is the mass, and ε is a scalar length small compared to m^{-1}.

We now calculate the energy of a free particle in the usual way, starting from (129). However, to avoid losing sight of (132), we must not perform coordinate integrations until *after* we perform the momentum integrations (the reverse of the usual procedure). The calculation leads to an increase in energy due to self-action by

$$\frac{\Delta m}{m} = -\frac{3e^2}{2\pi}\ln(m\varepsilon). \quad (134)$$

If we identify ε with the gravitational radius $2Gm$ of the particle (the smallest known length), we get

$$\frac{\Delta m}{m} = -\frac{3e^2}{2\pi}\ln(2Gm^2). \tag{135}$$

This is a finite correction, of the order $\Delta m/m \sim \frac{1}{3}$. The theory is thus capable of giving unambiguous finite answers, but the exact nature of ε remains to be understood.

The same ε occurs when we deal with closed loops. We have already seen how closed loops arise in the theory. In calculating a physical process, we must also include interactions with closed-loop paths in our influence functionals. This inclusion leads to a modification of effective charge in the usual way. The calculation is also similar to the usual one, but we must again perform the momentum integrals first and the coordinate integrals later, and we must keep the two legs of the delta functions apart by the rule (132). This procedure leads to a charge modification by

$$\frac{\Delta e}{e} = +\frac{2e^2}{3\pi}\ln(m\varepsilon). \tag{136}$$

With $\varepsilon = 2Gm$, $\Delta e/e \sim -\frac{1}{7}$.

We also note that the quadratic divergence of vacuum polarization does not appear at all. This divergence is due in the usual theory to the constant part of the electromagnetic potential. But, as we see from (107), if B_i is constant

$$\oint_{\Gamma^0} B_i\, dl^i = 0, \tag{137}$$

and no effect is produced by such a potential. This illustrates the advantage of being able to write the theory in a closed form.

To summarize, by quantizing the direct-particle theory, we have shown it to be as comprehensive as the Maxwell theory. That all scattering and decay processes are contained in the influence functionals discussed here shows the power of the path-integral technique. The phenomena of spontaneous transitions and decays of states are seen to be related strongly to cosmology and thermodynamics and to throw light on the nature of the arrow of time. These aspects of the theory are an advantage over Maxwell's theory. Finally, radiative corrections can be made finite by a rule that is intuitively easier to understand than the usual techniques of renormalization.

5

DIRECT-PARTICLE THEORIES IN RIEMANNIAN SPACE-TIME

§5.1. THE FOKKER ACTION (LECTURE 12)

We now return to the question, raised already in Chapter 3, of how to generalize the Fokker action to curved spacetime. In this chapter we shall also consider certain general aspects of direct-particle theories.

We begin with the concept of parallel propagators. Suppose two points A and B in Riemannian space are connected by a unique geodesic Γ_{AB}. Let X^{i_A} be a 4-vector at point A. Since we are concerned with nonlocal problems, we use A as a subscript on the vector index i of X to show that X^{i_A} transforms as a vector at A. Suppose we move X^{i_A} parallel to itself along Γ_{AB}, so that when it arrives at B it transforms as a vector X^{i_B} at B. Since parallel propagation preserves linearity, we can write

$$X^{i_B} = \bar{g}^{i_B}{}_{i_A} X^{i_A}, \tag{1}$$

where the functions $\bar{g}^{i_B}{}_{i_A}$ depend on A, B, and on the geometry of spacetime. The index i_A can be raised with the metric tensor g^{ik} at A and the index i_B lowered with g_{ik} at B. Thus we have the quantities

$$\bar{g}^{i_B}{}_{i_A}, \qquad \bar{g}^{i_B i_A}, \qquad \bar{g}_{i_B i_A}, \tag{2}$$

determining parallel propagation of covariant and contravariant vectors. There is symmetry with respect to A and B,

$$\bar{g}_{i_A i_B} = \bar{g}_{i_B i_A}, \qquad \bar{g}^{i_A i_B} = \bar{g}^{i_B i_A}. \tag{3}$$

The functions (2) are called parallel propagators. Since they transform as vectors at A and B, such quantities are called "bivectors" or "two-point" vectors. We can similarly have biscalars and bitensors, etc.

Because the proper distance s_{AB} along Γ_{AB},

$$s_{AB} = \int_{\Gamma_{AB}} \sqrt{g_{ik}\, dx^i\, dx^k}, \tag{4}$$

is a biscalar, we can give a meaning to $\delta(s^2{}_{AB})$ in curved spacetime. Thus it might appear at first sight as if the Fokker action could be generalized by replacing the flat-space form

$$\delta(s^2{}_{AB})\eta_{ik}\, da^i\, db^k, \tag{5}$$

by

$$\delta(s^2{}_{AB})\bar{g}_{i_A k_B}\, da^{i_A}\, db^{k_B}. \tag{6}$$

Although this is certainly a possible procedure, it does not lead to the equations of electrodynamics. The required generalization comes from the curved-space version of the wave equation

$$\Box_A \delta(s^2{}_{AB}) = 4\pi\delta_4(A, B). \tag{7}$$

Indeed, we may regard $\delta(s^2{}_{AB})$ in (5) as the Green's function of the flat-space form of the scalar-wave equation

$$\Box\phi = 0. \tag{8}$$

In curved space, we again look for the Green's function of (8). We shall also be concerned with the Green's function of the vector-wave equation

$$\Box A_i + R_i{}^k A_k = 0. \tag{9}$$

Letting X be a variable point with respect to which derivatives are taken, we consider the scalar Green's function $\bar{G}(X, B)$ and the vector Green's function $\bar{G}_{i_X i_B}$ as the solutions of the equations

$$\Box\bar{G}(X, B) = [-\bar{g}(X, B)]^{-1/2}\, \delta_4(X, B), \tag{10}$$

$$\Box\bar{G}_{i_X i_B} + R_{i_X}{}^{k_X}\bar{G}_{k_X i_B} = [-\bar{g}(X, B)]^{-1/2}\bar{g}_{i_X i_B}\,\delta_4(X, B), \tag{11}$$

where $\bar{g}(X, B)$ is the determinant of the matrix $\bar{g}_{i_X i_B}$. Since

$$\lim_{X\to B} \bar{g}_{i_X k_B} = g_{ik}(B), \tag{12}$$

the delta functions $\delta_4(X, B)$ in (10), (11) permit $\bar{g}$ to be replaced by g. However, the writing of $\bar{g}$ and $\bar{g}_{i_X i_B}$ in (10) and (11) serves to remind us that we are dealing with two-point quantities. We further choose the solution of (10) or (11) to be symmetric with respect to X and B:

$$\bar{G}(X, B) = \bar{G}(B, X), \qquad \bar{G}_{i_X i_B} = \bar{G}_{i_B i_X}. \tag{13}$$

These Green's functions have been studied by various authors for other reasons (e.g., DeWitt and Brehme, 1960). We shall discuss briefly those properties which concern us directly.

First, we have a relation between $\bar{G}(X, B)$ and $\bar{G}_{i_X i_B}$:

$$[\bar{G}_{i_X i_B}]^{;i_X} = -\bar{G}(X, B)_{;i_B}. \tag{14}$$

This can be proved without great difficulty. We next write the formal solutions of (10) and (11) obtained by Hadamard's method of series expansion:

$$\bar{G}(X, B) = \frac{1}{4\pi}[\Delta^{1/2}\,\delta(s^2{}_{XB}) - \bar{v}(X, B)\,\theta(s^2{}_{XB})], \tag{15}$$

$$\bar{G}_{i_X i_B} = \frac{1}{4\pi}[\Delta^{1/2}\bar{g}_{i_X i_B}\,\delta(s^2{}_{XB}) - \bar{v}_{i_X i_B}\,\theta(s^2{}_{XB})], \tag{16}$$

where

$$\Delta = \det\left(\frac{1}{2}\,s^2{}_{XB;i_X i_B}\right)/\bar{g}(X, B). \tag{17}$$

The presence of the heaviside functions in (15) and (16) shows that if we have a delta-function source at B, its effect propagates not only along the null cone from B, but also *inside* it. This breakdown of Huygen's principle has no analogue in flat space. In flat space $\bar{v} = 0$, $\bar{v}_{i_X i_B} = 0$, $\Delta = 1$. If we go back to (6), we now see that it forms the leading part of $4\pi\bar{G}_{i_A i_B}$, but not the whole of it.

We are now in a position to generalize the Fokker action to

$$S = -\sum_a \int m_a \, da - \sum_{a<b}\sum 4\pi e_a e_b \int\int \bar{G}_{i_A i_B} \, da^{i_A} \, db^{i_B}. \tag{18}$$

The potentials and fields are defined by

$$A_{i_X}{}^{(b)} = 4\pi e_b \int \bar{G}_{i_X i_B} \, db^{i_B}, \qquad F_{ik}{}^{(b)} = A_{k;\,i}{}^{(b)} - A_{i;\,k}{}^{(b)}. \tag{19}$$

In view of (11) we have

$$\Box A_i{}^{(b)} + R_i{}^k A_k{}^{(b)} = 4\pi J_i{}^{(b)}, \tag{20}$$

where

$$J_{i_X}{}^{(b)}(X) = e_b \int \bar{g}_{i_X i_B} \frac{\delta_4(X, B) \, db^{i_B}}{[-\bar{g}(X, B)]^{1/2}} \tag{21}$$

has the required vector property of the current. Because of (14) we have

$$A_{i_X}{}^{(b);\,i_X} = -4\pi e_b \int \bar{G}(X, B)_{;i_B} \, db^{i_B} = 4\pi e_b[\bar{G}(X, B_1) - \bar{G}(X, B_2)]. \tag{22}$$

Here we have assumed the world line of b to be between points B_1 and B_2. If $B_1 \to -\infty$, $B_2 \to +\infty$ (where $-\infty$ means infinite past and $+\infty$ means infinite future), we get zero on the righthand side of (22) for all the usual cosmological spaces. In such spaces the gauge condition is satisfied unless there is creation of charge. Henceforth we shall restrict attention to these spaces and assume no net charge creation, so that

$$A^{(b)}{}_i{}^{;i} = 0, \qquad F^{(b)ik}{}_{;k} = -4\pi J^{(b)i}. \tag{23}$$

We have therefore generalized the Fokker action to Riemannian space-time and in a way that preserves conformal invariance. Consider a conformal transformation

$$g^*{}_{ik} = e^{2\zeta} g_{ik}, \tag{24}$$

and define

$$\left.\begin{aligned} A_{(a)i} &= \sum_{b \neq a} A^{(b)}{}_i, \\ A^*{}_{(a)i} &= A_{(a)i} + \phi_{(a);\,i}, \end{aligned}\right\} \tag{25}$$

where $\phi_{(a)}$ satisfies the scalar equation

$$\Box\phi_{(a)} + 2\zeta_{;i}[A_{(a)}{}^{i} + \phi_{(a)}{}^{;i}] = 0 \tag{26}$$

with respect to the matrix g_{ik}. It is not hard to show that

$$A^{*}{}_{(a)i}{}^{;i} = 0, \tag{27}$$

and hence that $A^{*}{}_{(a)i}$ satisfies the starred form of the wave equation (20). The derivative in (27) is calculated with respect to the metric $g^{*}{}_{ik}$, as are the derivatives in the wave equation for $A^{*}{}_{(a)i}$. Yet the ϕ-functions in (25) do not affect $F^{*}{}_{ik}$, which remains equal to F_{ik}. Nor do the ϕ-functions affect the electromagnetic part of the action,

$$\sum_{a} \frac{1}{2} e_a \int A^{*}{}_{(a)i}\, da^{i} = \sum_{a} \frac{1}{2} e_a \int A_{(a)i}\, da^{i}. \tag{28}$$

The physical content of the theory remains invariant with respect to (24).

§5.2. GRAVITATIONAL INFLUENCE OF DIRECT-PARTICLE FIELDS (LECTURE 13)

Do direct-particle fields exert any gravitational attraction, as ordinary fields do? We shall discuss this question for the explicit case of the electromagnetic field. The direct-particle formulation of electromagnetism has only the single action term

$$-\frac{1}{2} \sum_{a} \sum_{b} 4\pi e_a e_b \int\int \bar{G}_{i_A i_B}\, da^{i_A}\, db^{i_B}, \tag{29}$$

whereas the Maxwell action has two terms

$$-\frac{1}{16\pi} \int F_{lm} F^{lm} \sqrt{-g}\, d^4x - \sum_{a} e_a \int A_i\, da^{i}. \tag{30}$$

The form (29) differs from the action used in Chapter 3, not only in having $4\pi\bar{G}_{i_A i_B}$ in place of $\delta(s^2{}_{AB})$, but also in having unrestricted summations with respect to a, b. Thus in (29) we can have $a = b$. From what was said in Chapter 4 concerning self-action, it might seem that in classical theory, with timelike paths for the particles, the self-terms in (29) would be zero provided points A, B, were never permitted to coincide. This was the case in flat space. However, it is possible for $\bar{G}_{i_A i_B}$ to be nonzero in curved space even though A, B, are distinct points on the same timelike path, because of the $\bar{v}_{i_A i_B}$ function which appears in $\bar{G}_{i_A i_B}$ [see equation (16)]. Hence the omission or inclusion of the self-terms is a more acute question in curved spacetime than it is in flat space. Since the self-terms are needed in quantum electrodynamics, it is correct to include them also in (29), even though the discussion is classical. To avoid infinities, points A and B in such terms must never be allowed to coincide.

The gravitational effect of a field or of an interaction is expressed by an energy-momentum tensor, T^{ik}, say, defined by expressing the change of the action corresponding to a variation of g_{ik},

$$g_{ik} \to g_{ik} + \delta g_{ik}, \tag{31}$$

in the form

$$-\frac{1}{2}\int_V T^{ik}\, \delta g_{ik} \sqrt{-g}\, d^4x, \tag{32}$$

V being the four-dimensional volume in which δg_{ik} is nonzero. Thus to obtain the electromagnetic energy-momentum tensor $T^{ik}{}_{(\mathrm{em})}$ we have to calculate the variation of (29) for the direct-particle theory, and of (30) for the Maxwell theory. The result in the latter case is well-known, and is

$$T^{ik}{}_{(\mathrm{em})} = \frac{1}{4\pi}\left(\frac{1}{4} g^{ik} F^{lm} F_{lm} - F^{il} F^{k}{}_{l}\right), \tag{33}$$

in which the electromagnetic field tensor F_{lm} is taken to be retarded.

Because neither A_i nor da^i changes when g_{ik} is varied, the second term of (30) does not contribute to $T^{ik}{}_{(\mathrm{em})}$ in the Maxwell theory. Thus (33) is derived entirely from the first term of (30). Since (29) has no analogue to this first term, but only one to the second term of (30), it might seem at first sight as if the direct-particle theory would lead to an energy-momentum tensor that was identically zero. Yet it does not, because a change of geometry changes the solutions of the wave equations (10) and (11). Consequently $\bar{G}_{i_A i_B}$ in (29) changes. To calculate this effect, define

$$\mathcal{F}_{i_X k_X i_B} = \frac{\partial \bar{G}_{k_X i_B}}{\partial x^{i_X}} - \frac{\partial \bar{G}_{i_X i_B}}{\partial x^{k_X}}. \tag{34}$$

The electromagnetic field generated by particle b can be written as

$$F^{(b)}_{i_X k_X} = 4\pi e_b \int \mathcal{F}_{i_X k_X i_B}\, db^{i_B}, \tag{35}$$

and the wave equation (11) can be written in the following way:

$$\begin{aligned}\sqrt{-g}(g^{il}g^{mk}\bar{G}_{li_B;mk} + R^{im}\bar{G}_{mi_B})\\ &\equiv \frac{\partial}{\partial x^k}(g^{il}g^{mk}\sqrt{-g}\mathcal{F}_{mli_B}) + \sqrt{-g}\,g^{il}\bar{G}^k{}_{i_B;kl}\\ &= \delta_4(X, B)\bar{g}^i{}_{i_B}. \qquad (36)\end{aligned}$$

For simplicity, the subscript X on indices taken at X has been suppressed in (36).

The change (31) does not alter $\delta_4(X, B)\bar{g}^i{}_{i_B}$, while

$$\begin{aligned}\delta\left[\frac{\partial}{\partial x^k}(g^{il}g^{mk}\sqrt{-g}\mathcal{F}_{mli_B})\right] &\equiv \frac{\partial}{\partial x^k}(g^{il}g^{mk}\sqrt{-g}\,\delta\mathcal{F}_{mli_B})\\ &\quad + \frac{\partial}{\partial x^k}[\mathcal{F}_{mli_B}\,\delta(g^{il}g^{mk}\sqrt{-g})], \qquad (37)\end{aligned}$$

$$\begin{aligned}\delta(g^{il}\sqrt{-g}\,\bar{G}^k{}_{i_B;kl}) &= \delta\left[g^{il}\sqrt{-g}\frac{\partial}{\partial x^l}(\bar{G}^k{}_{i_B;k})\right]\\ &= \delta(g^{il}\sqrt{-g})\frac{\partial}{\partial x^l}(\bar{G}^k{}_{i_B;k})\\ &\quad + \sqrt{-g}\,g^{il}\frac{\partial}{\partial x^l}\left[\delta\left(\frac{1}{\sqrt{-g}}\right)\frac{\partial}{\partial x^k}(\sqrt{-g}\,\bar{G}^k{}_{i_B})\right]\\ &\quad + \sqrt{-g}\,g^{il}\frac{\partial}{\partial x^l}\left\{\frac{1}{\sqrt{-g}}\frac{\partial}{\partial x^k}[\delta(\sqrt{-g}\,g^{km})\bar{G}_{mi_B}]\right\}\\ &\quad + \sqrt{-g}\,g^{il}\frac{\partial}{\partial x^l}\left[\frac{1}{\sqrt{-g}}\frac{\partial}{\partial x^k}(\sqrt{-g}\,g^{km}\,\delta\bar{G}_{mi_B})\right], \qquad (38)\end{aligned}$$

so that the variation of (36) gives

$$\begin{aligned}\frac{\partial}{\partial x^k}(g^{il}g^{mk}\sqrt{-g}\,\delta\mathcal{F}_{mli_B}) + \sqrt{-g}\,g^{il}\frac{\partial}{\partial x^l}\left[\frac{1}{\sqrt{-g}}\frac{\partial}{\partial x^k}(\sqrt{-g}\,g^{km}\,\delta\bar{G}_{mi_B})\right]\\ = -\frac{\partial}{\partial x^k}[\delta(g^{il}g^{mk}\sqrt{-g})\mathcal{F}_{mli_B}] - \delta(\sqrt{-g}\,g^{il})\frac{\partial}{\partial x^l}(\bar{G}^k{}_{i_B;k})\\ - \sqrt{-g}\,g^{il}\frac{\partial}{\partial x^l}\left[\delta\left(\frac{1}{\sqrt{-g}}\right)\frac{\partial}{\partial x^k}(\sqrt{-g}\,\bar{G}^k{}_{i_B})\right]\\ - \sqrt{-g}\,g^{il}\frac{\partial}{\partial x^l}\left\{\frac{1}{\sqrt{-g}}\frac{\partial}{\partial x^k}[\delta(\sqrt{-g}\,g^{km})\bar{G}_{mi_B}]\right\}. \qquad (39)\end{aligned}$$

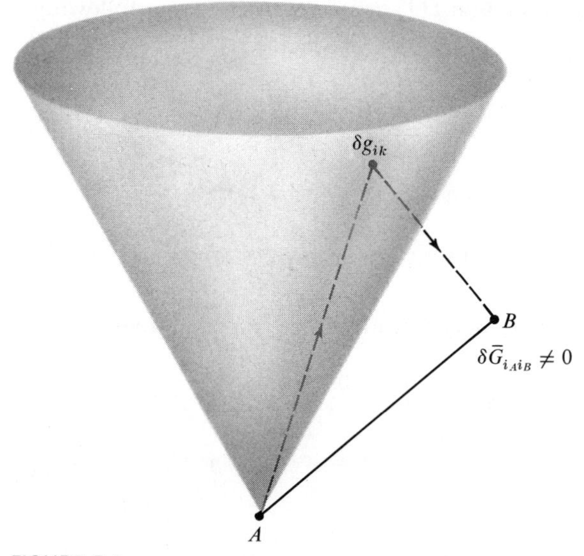

FIGURE 5.1
The requirement that $\delta \bar{G}_{i_A i_B}$ be zero, with respect to $g_{ik} \to g_{ik} + \delta g_{ik}$, when B lies appreciably outside the light cone from A, leads to the canonical electromagnetic tensor. If scattering backward from points within the light cone were permitted, then $\delta \bar{G}_{i_A i_B}$ would be nonzero, and we would obtain the so-called Frenkel tensor.

The differential operator acting on $\delta \bar{G}_{mi_B}$ on the lefthand side of (39) is the same as that acting on $\bar{G}_{mi_B}$ in (36). We can therefore use appropriate solutions of the wave equation (11) to obtain $\bar{G}_{i_A i_B}$. Some care is needed, however, in handling the retarded and advanced parts of the source term in (39). Restoring the subscript X, and writing $\delta Q^{i_X}{}_{i_B}$ for the righthand side of (39), we have

$$Q^{i_X}{}_{i_B} = Q^{i_X}{}_{i_B}{}^{(\text{ret})} + Q^{i_X}{}_{i_B}{}^{(\text{adv})}, \tag{40}$$

where the retarded and advanced parts in (40) are obtained by putting

$$\bar{G}_{i_X i_B} = \frac{1}{2}[G_{i_X i_B}{}^{(\text{ret})} + G_{i_X i_B}{}^{(\text{adv})}] \tag{41}$$

in $Q^{i_X}{}_{i_B}$; $G_{i_X i_B}{}^{(\text{ret})}$, $G_{i_X i_B}{}^{(\text{adv})}$, being the retarded and advanced solutions of (11). The issue requiring particular care is that the correct propagator for $Q^{i_X}{}_{i_B}{}^{(\text{ret})}$ is $G_{i_A i_X}{}^{(\text{ret})}$, and that for $Q^{i_X}{}_{i_B}{}^{(\text{adv})}$ is $G_{i_A i_X}{}^{(\text{adv})}$. Thus

$$\delta \bar{G}_{i_A i_B} = \int_V [G_{i_A i_X}{}^{(\text{ret})} Q^{i_X}{}_{i_B}{}^{(\text{ret})} + G_{i_A i_X}{}^{(\text{adv})} Q^{i_X}{}_{i_B}{}^{(\text{adv})}]\, d^4x. \tag{42}$$

The reason for this separation of the retarded and advanced parts of $Q^{i_X}{}_{i_B}$ will be understood from Figure 5.1. Since

$$G_{i_A i_X}{}^{(\text{ret})} = G_{i_X i_A}{}^{(\text{adv})}, \qquad G_{i_A i_X}{}^{(\text{adv})} = G_{i_X i_A}{}^{(\text{ret})}, \tag{43}$$

we can write, equivalently,

$$\delta \bar{G}_{i_A i_B} = \int_V [G_{i_X i_A}{}^{(\text{adv})} Q^{i_X}{}_{i_B}{}^{(\text{ret})} + G^{(\text{ret})}{}_{i_X i_A} Q^{i_X}{}_{i_B}{}^{(\text{adv})}]\, d^4x. \tag{44}$$

The two terms in (44) are treated similarly; so it is sufficient to consider one of them. As usual, we set the δg_{ik} and their first derivatives at zero on the boundary of V. Defining

$$\mathscr{F}_{mli_B}{}^{(\text{ret})} = \frac{\partial G_{li_B}{}^{(\text{ret})}}{\partial x^m} - \frac{\partial G_{mi_B}{}^{(\text{ret})}}{\partial x^l},$$

$$\mathscr{F}_{mli_B}{}^{(\text{adv})} = \frac{\partial G_{li_B}{}^{(\text{adv})}}{\partial x^m} - \frac{\partial G_{mi_B}{}^{(\text{adv})}}{\partial x^l}, \tag{45}$$

and using the divergence theorem, it is not hard to see that

$$\begin{aligned} 2\int_V G_{i_X i_A}{}^{(\text{adv})} Q^{i_X}{}_{i_B}{}^{(\text{ret})}\, d^4x = &-\frac{1}{2}\int_V \delta(g^{il} g^{mk}\sqrt{-g})\mathscr{F}_{iki_A}{}^{(\text{adv})}\mathscr{F}_{mli_B}{}^{(\text{ret})}\, d^4x \\ &- \int_V \delta(\sqrt{-g}\, g^{il}) G_{ii_A}{}^{(\text{adv})} \frac{\partial}{\partial x^l}[G^{(\text{ret})k}{}_{i_B;k}]\, d^4x \\ &- \int_V \delta(\sqrt{-g}\, g^{il}) \frac{\partial}{\partial x^l}[G^{(\text{adv})k}{}_{i_A;k}] G_{ii_B}{}^{(\text{ret})}\, d^4x \\ &- \int \delta(\sqrt{-g}) G^{(\text{adv})k}{}_{i_A;k} G^{(\text{ret})l}{}_{i_B;l}\, d^4x. \end{aligned} \tag{46}$$

The index X has again been suppressed on the righthand side of (46) from indices taken at point X.

To calculate the variation of (29), we simply integrate along the world lines of a and b, and take the double sum with respect to a, b. If there is no net charge creation we get, as explained before,

$$\int \bar{G}^l{}_{i_B;l}\, db^{i_B} = -\int \bar{G}_{;i_B}\, db^{i_B} = 0. \tag{47}$$

Since the advanced and retarded parts of $\int \bar{G}_{;i_B}\, db^{i_B}$ vanish separately, only the first term of (46) survives. Moreover, after the sums with respect to a, b, this term becomes symmetric in the advanced and retarded fields. The two terms of (44) are therefore seen to give equal contributions. Because of the factor 2 on the lefthand side, it follows that (46) gives the full contribution of (44). Using (35) in the form

$$F^{\text{ret}(b)}{}_{ik} = 4\pi e_b \int \mathcal{G}_{iki_B}{}^{(\text{ret})}\, db^{i_B}, \; F^{\text{adv}(b)}{}_{ik} = 4\pi e_b \int \mathcal{G}_{iki_B}{}^{(\text{adv})}\, db^{i_B}, \quad (48)$$

we have

$$-\frac{1}{2}\delta \sum_a \sum_b 4\pi e_a e_b \int\int \bar{G}_{i_A i_B}\, da^{i_A}\, db^{i_B}$$
$$= \frac{1}{16\pi} \sum_a \sum_b \int_V \delta(g^{il} g^{mk} \sqrt{-g}) F^{\text{ret}(b)}{}_{ml} F^{\text{adv}(a)}{}_{ik}\, d^4x. \quad (49)$$

Remembering the definition (32) of the energy-momentum tensor, we find

$$T^{ik}{}_{(\text{em})} = \frac{1}{4\pi}\left\{\frac{1}{4} g^{ik} \left[\sum_a F^{\text{adv}(a)}{}_{lm}\right]\left[\sum_b F^{\text{ret}(b)lm}\right] - \left[\sum_a F^{\text{adv}(a)il}\right]\left[\sum_b F^{\text{ret}(b)k}{}_l\right]\right\}. \quad (50)$$

The result (50) is similar to the energy-momentum tensor of the Maxwell theory, except that the advanced and retarded fields appear symmetrically in (50), whereas only retarded fields occur in the Maxwell theory. However, (50) can easily be reduced to the usual form when account is taken of the response of the universe.

Consider the slab of spacetime $0 \leq t \leq T$. For any point X in the slab we have

$$F^{\text{ret}(a)} = F^{\text{ret}(a)}{}_{0 \leq t \leq T} + F^{\text{ret}(a)}{}_{t < 0},$$
$$F^{\text{adv}(a)} = F^{\text{adv}(a)}{}_{0 \leq t \leq T} + F^{\text{adv}(a)}{}_{t > T}, \quad (51)$$

for any particle a. As in Chapter 4, we consider the case $F^{\text{ret}}{}_{t<0} = 0$, and express the response condition in the form

$$\sum_a F^{\text{adv}(a)}{}_{t>T} = \sum_a [F^{\text{ret}(a)}{}_{0 \leq t \leq T} - F^{\text{adv}(a)}{}_{0 \leq t \leq T}]. \quad (52)$$

Substituting (52) in (50) leads immediately to

$$T^{ik}{}_{(\mathrm{em})} = \frac{1}{4\pi}\left\{\frac{1}{4}g^{ik}\left[\sum_a F_{lm}{}^{\mathrm{ret}(a)}\right]\left[\sum_b F^{\mathrm{ret}(b)lm}\right] - \left[\sum_a F^{\mathrm{ret}(a)il}\right]\left[\sum_b F^{\mathrm{ret}(b)k}{}_l\right]\right\}, \quad (53)$$

which is the same as the usual energy-momentum tensor.

§5.3. A GENERAL CORRESPONDENCE BETWEEN FIELD THEORIES AND DIRECT-PARTICLE THEORIES

The emergence of an energy tensor in the above calculation and its similarity to the energy tensor (33) of field theory is surprising at first. But this is no accident. Rather, there exists a general correspondence between the field concept and the action-at-a-distance concept, of which electromagnetism is a special case. We now state a general result obtained by Narlikar (1968) that covers a wide range of cases.

Let ϕ be a tensor field of rank N in interaction with particles and given by an action of the type

$$\int L[\phi]\sqrt{-g}\, d^4x + \sum_a \int I[\phi, a]\, da. \quad (54)$$

The first term contains a field Lagrangian, the second describes the interaction of ϕ with particles. $L[\phi]$ and $I[\phi, a]$ satisfy the following conditions.

First, $L[\phi]$ is a bilinear invariant composed of ϕ and of its first derivatives. The coefficients appearing in the invariant are functions of the spacetime geometry. Thus for a tensor field ϕ_{ik}, $L[\phi]$ would be a combination of the form

$$A^{kl}{}_{mn}\phi_{lk}\phi^{mn} + B^{ikl}{}_{nmp}\phi_{ik;l}\phi^{mn;p} + C^{ikl}{}_{mn}\phi_{ik;l}\phi^{mn}, \quad (55)$$

where A, B, C, are tensors connected with the spacetime geometry.

Second, $I[\phi, a]$ is an invariant expression of the form

$$I[\phi, a] = D\phi \cdot \xi^{(a)}, \quad (56)$$

where D is a coupling constant and $\xi^{(a)}$ is a tensor of rank N depending only on the world line of particle a.

To a field theory given by (54), there corresponds a direct-particle theory of the form

$$\frac{1}{2}\sum_{a}\sum_{b} D^2 \int\int \xi^{(a)} \cdot \bar{G}_{AB} \cdot \xi^{(b)}\, da\, db, \tag{57}$$

where $\bar{G}_{AB}$ is the symmetric Green's function of the field equation satisfied by ϕ. Thus $\bar{G}_{AB}$ is a tensor of rank N at each end A, B. The direct field due to particle b is given by

$$\phi^{(b)}(X) = D \int \bar{G}_{XB} \cdot \xi^{(b)}\, db. \tag{58}$$

It can be shown that the variation of geometry expressed by (31) leads to similar energy-momentum tensors in the two theories.

We end this chapter by giving the field theory corresponding to the inertia theory described later. The mass field $m(X)$ has the action given by

$$-\int \left(m_i m^i - \frac{1}{6} R m^2\right) \sqrt{-g}\, d^4x - \sum_{a} \int m\, da, \tag{59}$$

where $m_i = \partial m / \partial x^i$. It is easily verified that (59) is conformally invariant. A theory of this type has recently been used by particle physicists to introduce scale invariance, the massless particles given by

$$\Box m + \frac{1}{6} R m = 0 \tag{60}$$

being called "dilatons."

6

INERTIA AND GRAVITATION TREATED CLASSICALLY

§6.1. UNITS IN PHYSICS (LECTURE 14)

Let us return for a while to the flat space of special relativity. A general point X in spacetime is described by four coordinates t, $\boldsymbol{x}$, with the 3-vector $\boldsymbol{x} = (x, y, z)$ carrying the spatial coordinates x, y, z. The origin is the point $t = 0, \boldsymbol{x} = 0$, and the square of the four-dimensional distance from the origin to X is given by

$$t^2 - \boldsymbol{x}^2. \tag{1}$$

Different observers in uniform relative motion with respect to each other can all choose the same spacetime point as origin, but then they find that the coordinate values which they attach to the general point X are not usually the same. However, (1) calculated for any X by each observer is the same.

Sometimes the velocity of light c is included in (1), $c^2t^2 - \boldsymbol{x}^2$. This is necessary if we elect to use a unit for measuring time that is different from the unit used for measuring spatial intervals. Although such a procedure may be convenient in some practical applications, there is no more logical reason for using a different time unit than there would be for measuring x, y, z, in different units. Apart from the sign difference in (1), all four coordinates appear on the same footing, and should be described in terms of the same unit. Then $c = 1$, all particle velocities have magnitudes between 0 and 1, and these magnitudes are dimensionless.

In the MKS and SI unit systems, it is recommended that x, y, z, be measured in meters and t in seconds. With the definitions used for the meter and the second, one then has

$$c = 299{,}792{,}500 \text{ meters second}^{-1}. \tag{2}$$

This is said to be the "velocity of light," but the velocity of light is unity; so (2) only describes the relation between "meter" and "second,"

$$1 \text{ second} \equiv 299{,}792{,}500 \text{ meters}. \tag{3}$$

The number 299,792,500 is manmade.

In the cgs unit system, the relation between the spatial derivatives of the magnetic vector $\boldsymbol{H}$ and the time derivatives of the electric vector $\boldsymbol{E}$ is

$$\operatorname{curl} \boldsymbol{H} = \frac{4\pi}{c}\boldsymbol{j} + \frac{1}{c}\frac{\partial \boldsymbol{E}}{\partial t}. \tag{4}$$

The scale factor $c = 3 \cdot 10^{10}$ cm sec^{-1} (telling us that one second is the same as $3 \cdot 10^{10}$ centimeters) belongs with t in the second term on the righthand side, $\partial \boldsymbol{E}/\partial(ct)$, correctly reminding us that the same unit must be used for taking both the time and space derivatives. And the current vector $\boldsymbol{j}$ is determined by the motion of charged particles. In the cgs system, the velocity vectors of the particles have to be scaled by c^{-1}. This is why c^{-1} also appears in the first term on the righthand side of (4).

The cgs system is logically defective in using different units for t and $\boldsymbol{x}$, but at least it always reminds us where scale factor corrections must be introduced. This reminding permits calculations to be made without serious risk of error, and without distorting the meaning of the physical equations themselves. In the MKS and SI systems, on the other hand, the scale factors c^{-1} in (4) are absorbed into $\boldsymbol{j}$ and $\boldsymbol{E}$. Hence not only are t and $\boldsymbol{x}$ scaled differently, but there is also a further difference between $\boldsymbol{E}$ and $\boldsymbol{H}$, which should properly fit together to form the six components of the antisymmetric field tensor F_{ik} and therefore should *not* be differently scaled. These latter systems therefore lead to confused thinking.

Further, there is no advantage to be gained from attempting to suppress the 4π in (4). The unit sphere in Euclidean 3-space has area 4π. However intensely the protagonists of modern unit systems may regret this fact, it happens to be true, and it causes 4π to appear in various physical equations.

If one suppresses the 4π factor in one place, it appears inevitably somewhere else.

The above discussion contains a still undefined concept, that an observer is able to measure coordinate differences $t_2 - t_1$, $\boldsymbol{x}_2 - \boldsymbol{x}_1$, between points X_1, X_2, in terms of some length unit L. We must consider next how this is to be done. Spacetime itself provides no answer. Spacetime is not divided up like a four-dimensional piece of graph paper into equally spaced intervals L that serve as a grid. We must appeal to physics, not to geometry.

A plane electromagnetic wave of frequency k varies with position X in a region free from particles according to the phase factor

$$\exp[i(\boldsymbol{k}\cdot\boldsymbol{x} - kt)], \qquad k = |\boldsymbol{k}|, \tag{5}$$

the 3-vector $\boldsymbol{k}$ giving the direction of propagation of the wave. Such a physical disturbance can be used to establish a spacetime grid. For a specific spatial coordinate, say, x, choose the wave with $\boldsymbol{k}$ in the direction of the x-axis. Then (5) takes the form $\exp[ik(x - t)]$. The grid can be imagined to be set up in the following way. On any hyperplane $t =$ constant, the phase factor has periodicity $2\pi k^{-1}$ with respect to x, and on any plane $x =$ constant the phase factor has the same periodicity with respect to the time-coordinate. Similar periodicities can also be set up for the y, z, coordinates. A standard grid can thus be obtained using the k^{-1} value appropriate to the radiation emitted by an explicit atomic transition. The grid has the desirable property that it gives the same unit for the t coordinate as for the spatial coordinates. However, this important property is unfortunately obscured in practice because different atomic transitions are used for time and for space. For the space coordinates, the $2p_{10}$ to $5d_5$ transition of ^{86}Kr is used, whereas for the time-coordinate the transition used is that between the two hyperfine levels of the ground state of ^{133}Cs.

The merit of using an explicit transition to determine a specific k^{-1} value lies in the fact that, if we choose a suitably long-lived excited atomic level, the k^{-1} value can be well-defined. Such a choice means, however, that our basic length unit will be determined by a somewhat obscure transition of a complicated atom. The k^{-1} value in question then depends on the electron mass, the fine-structure constant, and the nuclear properties of the chosen atom. A precise, direct calculation of such a k^{-1} value from the basic principles of atomic structure is beyond present-day methods, and to this extent the procedure remains intellectually unsatisfactory. One might hope to define a length unit L in a simpler and more elegant way. It may still happen, in those macroscopic problems where the greatest accuracy is needed, that it is preferable to work empirically in terms of the transitions of complex atoms, ^{86}Kr or ^{133}Cs, for example, but for most purposes, and especially for obtaining insight into the nature of physical dimensions and units, a simpler determination of L is still needed.

From here on we write

$$x, y, z, t \equiv x^1, x^2, x^3, x^4 \tag{6}$$

whenever we wish to use four-dimensional notation. Thus

$$x_i x^i = \eta_{ik} x^i x^k, \qquad i, k = 1, 2, 3, 4, \tag{7}$$

where $\eta_{ik} = \text{diag}(-1, -1, -1, +1)$ is the Minkowski tensor.

Consider a nonrelativistic classical particle that goes from point X_1 to point X_2. The path by which it does so can be determined from the action

$$S = \int_{X_1}^{X_2} \mathcal{L}\, dt, \tag{8}$$

where $\mathcal{L}$ is the Lagrangian and the integral is taken along a specific path. Suppose the path used in (8) to be subject to a small variation. Let $S + \delta S$ be the new value of the action. If δS is zero to the first order in small quantities for every such variation, then the path from which we started happens to be the classical path. To give expression to this procedure, one uses the usual methods of the calculus of variations, the resulting differential equations being the "equations of motion" of the particle.

For a nonrelativistic free particle of mass m, the Lagrangian is just the kinetic energy $\frac{1}{2} m\dot{\mathbf{x}}^2$. For a nonrelativistic particle in an electromagnetic field with 4-potential A_i, we have

$$\mathcal{L} = \frac{1}{2} m\dot{\mathbf{x}}^2 - eA_i \frac{dx^i}{dt}, \qquad i = 1, 2, 3, 4, \tag{9}$$

where e is the charge of the particle, and dx^i is a four-dimensional coordinate displacement along the path.

It is easy to pass to the relativistic case. Inserting (9) in (8),

$$S = \frac{1}{2} m \int_{X_1}^{X_2} \dot{\mathbf{x}}^2\, dt - e \int_{X_1}^{X_2} A_i\, dx^i, \tag{10}$$

and the electromagnetic term is immediately invariant. Let da be an element of four-dimensional length along the path, so that

$$da^2 = dt^2 - (d\mathbf{x})^2, \qquad da > 0. \tag{11}$$

Thus

$$da = dt[1 - \dot{x}^2]^{1/2} = dt\left[1 - \frac{1}{2}\dot{x}^2 + \ldots\right]. \tag{12}$$

Replacing (10) by

$$S = -m\int_{X_1}^{X_2} da - e\int_{X_1}^{X_2} A_i\, dx^i, \tag{13}$$

it is seen that the term in $\dot{x}^2$ in the expansion of da yields the $\dot{x}^2$ term in (10). There is also a term

$$-m\int_{X_1}^{X_2} dt = m(t_1 - t_2), \tag{14}$$

but this does not affect the calculation of the path, since it involves only m and the time-coordinates of X_1, X_2, which quantities do not change when the path is varied. The term of order $(\dot{x}^2)^2$ in (13) is neglected in the nonrelativistic approximation. The required relativistic form of S is given by (13).

When the equations of motion are determined by requiring $\delta S = 0$ with respect to small variations of the path, the electromagnetic term in (13) is found to have the effect of subjecting the particle to a force, usually referred to as the Lorentz force. It is given by

$$eF_{ik}\frac{dx^k}{da}, \qquad i, k = 1, 2, 3, 4, \tag{15}$$

where F_{ik} is defined by

$$F_{ik} = \frac{\partial A_k}{\partial x^i} - \frac{\partial A_i}{\partial x_k}. \tag{16}$$

The above formalism can readily be adapted to nonrelativistic quantum mechanics. Instead of seeking a unique path from X_1 to X_2, consider all paths without time-reversals, and evaluate

$$\sum_{\text{paths}} \exp\left(\frac{iS}{\hbar}\right). \tag{17}$$

Provided a suitable measure of paths is used in this summation, the result has the physical interpretation that the square of the modulus of (17) gives the probability that the particle, initially at X_1, goes to X_2.

The phase angle $S/\hbar$ in (17) must be dimensionless. Indeed, the constant $\hbar$ has been introduced to ensure that this is so. Here $\hbar = h/2\pi$, where h is the old form of Planck's constant. Instead of inserting $\hbar$ explicitly in (17), we could, however, have redefined the Lagrangian to be

$$\mathcal{L} = \frac{1}{2\hbar} m\dot{x}^2 - \frac{e}{\hbar} A_i \frac{dx^i}{dt}. \tag{18}$$

Instead of (17) we would then have

$$\sum_{\text{paths}} \exp(iS). \tag{19}$$

That is, the action S given by (8) with $\mathcal{L}$ containing the $\hbar^{-1}$ factor would be dimensionless, and the Lagrangian itself would be an inverse length, L^{-1}.

The dimensionality of the electromagnetic 4-potential A_i can readily be determined by considering the classical field. Denoting the particle by b, the contravariant form A^i of the 4-potential due to b at a general point X is given by

$$A^i(X) = e \int \delta(s^2{}_{XB})\, db^i, \tag{20}$$

where db^i is a coordinate displacement along the path of the particle, B is a point at db^i, $s^2{}_{XB}$ is the square of the four-dimensional distance between B and X, and $\delta(s^2{}_{XB})$ is the Dirac function with the property

$$\int_{-\infty}^{\infty} \delta(s^2)\, ds^2 = 1. \tag{21}$$

The covariant and contravariant forms of the electromagnetic potential are connected by

$$A_i = \eta_{ik} A^k. \tag{22}$$

From (21) it follows that $\delta(s^2)$ has dimensionality L^{-2}; so the righthand side of (20) has dimensionality eL^{-1}. Since A_i has the same dimensionality as A^i, it follows that the dimensionality of the electromagnetic term in (18) is $(e^2/\hbar)L^{-1}$, whence we see that $e^2/\hbar$ must be dimensionless. With e the elementary charge, $e^2/\hbar$ is the fine-structure constant, with the experimentally determined value 7.297351×10^{-3}. Since e^2 appears in association with $\hbar$ in the form (18) for $\mathcal{L}$, and since the properties of physical systems are determined from $\mathcal{L}$, it follows that the elementary charge will always appear in physics in association with $\hbar$, as $e^2/\hbar$. Hence there is nothing to stop $\hbar$ being absorbed into e^2, in which case the elementary charge is dimensionless, and satisfies

$$e^2 = 7.297351 \times 10^{-3}. \tag{23}$$

The form (18) of the Lagrangian for a nonrelativistic particle is now

$$\mathcal{L} = \frac{m}{2\hbar}\dot{x}^2 - eA_i\frac{dx^i}{dt}, \tag{24}$$

with the understanding that A_i is to be determined with the elementary charge satisfying (23). Can $\hbar$ also be absorbed into m? Since m will never appear except in association with $\hbar$, there is again no reason why we should not do so. This is for nonrelativistic quantum mechanics. The same situation occurs also in the relativistic theory. Writing γ^k for the Dirac matrices, $k = 1, 2, 3, 4$, the Dirac equation for a free particle of mass m is

$$\gamma^k\frac{\partial\psi}{\partial x^k} + \frac{im}{\hbar}\psi = 0, \tag{25}$$

and the mass occurs only in the form $m/\hbar$.

After absorbing $\hbar$ into m, it is clear that m must have dimensionality L^{-1}, since $\mathcal{L}$ given by (24) itself has dimensionality L^{-1}. A length unit can therefore be defined in terms of the mass of some chosen particle. Choosing the electron,

$$L = m_e^{-1} = C, \tag{26}$$

where C is the Compton wavelength of the electron. This choice of the electron as standard is useful in many problems, because $e^{-2}C$ determines the order of the particle separation in macroscopic bodies, and $(e^4/4\pi)C^{-1}$ is the Rydberg constant. Thus C and C^{-1}, together with appropriate powers of e^2, determine both the characteristic emitted frequencies from atomic transitions and the structure of macroscopic bodies.

Omitting gravitation, which involves a departure from flat space, physics is concerned with a set of particle masses and with three dimensionless coupling constants—in addition to e^2 for the electromagnetic field, there are dimensionless constants for the weak and strong interactions. From these ingredients, together with the field propagators [the function $\delta(s^2_{XB})$ in (20) was the electromagnetic propagator], all of physics (excepting gravitation) must be constructed. The field propagators have dimensionalities [$\delta(s^2_{XB})$ had dimensionality L^{-2}], but they do not determine a length scale. The only length scale available is that given by the mass of some reference particle, the masses of all other particles then being determined by dimensionless ratios to the mass of the reference particle. It follows that the procedure suggested here, of determining the length scale from the electron mass, is not just one of many possible procedures. Except insofar as some other particle might be chosen as standard, *it is the only possible procedure.*

All physical quantities now have the form

$$\text{(dimensionless number)} \cdot C^n, \tag{27}$$

the power n of C determining the dimensionality, and the dimensionless number giving the magnitude of the quantity. All physical measurements are concerned with determining such magnitudes. No laboratory measurement tells us about C itself. According to many standard texts,

$$2\pi C = 2.4263096 \times 10^{-12} \text{ meters}, \tag{28}$$

it being implied that (28) gives information about the electron. It doesn't! What (28) gives information about is the "meter," namely,

$$1 \text{ meter} = 2\pi/(2 \cdot 4263096 \times 10^{-12}) \cdot C. \tag{29}$$

§6.2. CONFORMAL INVARIANCE

Suppose an observer at point X receives radiation emitted by atoms at point X' that possess a characteristic pattern of spectrum lines (e.g., the Balmer series of hydrogen). Let λ be the measured wavelength of a particular line, and let λ_0 be the wavelength of the same line emitted by atoms at rest in the observer's laboratory at X. It is always found that

$$1 + z = \frac{\lambda}{\lambda_0} \tag{30}$$

is the same for all such lines. How are we to understand the possibility that z might not be zero?

One explanation for $z \neq 0$ is that the material at X' is in motion with respect to the rest frame of the observer at X. We then have the well-known Doppler effect determined in the following way.

The index $k \cdot x \equiv \eta_{lm} k^l x^m$ appearing in a plane electromagnetic wave is dimensionless and invariant with respect to a Lorentz transformation. Suppose the velocity of X' to be described by the spatial 3-vector $(v, 0, 0)$ with respect to axes x, y, z, at X. The Lorentz relation between coordinates $(\boldsymbol{x}, t)$ with respect to the observer at X and coordinates $(\boldsymbol{x}', t')$ with respect to the observer at X', the two observers being supposed coincident at $t = 0$, is

$$x = \frac{x' + vt'}{(1 - v^2)^{1/2}}, \qquad y = y', \qquad z = z', \qquad t = \frac{t' + vx'}{(1 - v^2)^{1/2}}. \tag{31}$$

In order that

$$k \cdot x \equiv k_x x + k_y y + k_z z - k_t t = k' \cdot x',$$

we require the contravariant 4-vectors (k_x, k_y, k_z, k_t), (k_x', k_y', k_z', k_t'), to be related by

$$k_x = \frac{k_x' + vk_t'}{(1 - v^2)^{1/2}}, \qquad k_y = k_y', \qquad k_z = k_z', \qquad k_t = \frac{k_t' + vk_x'}{(1 - v^2)^{1/2}}. \tag{32}$$

Writing

$$\cos\alpha = k_x/k_t, \qquad \cos\alpha' = k_x'/k_t', \tag{33}$$

we can express the first and last of the relations (32) in the form

$$k_x = k_t \cos\alpha = \frac{k_t'(\cos\alpha' + v)}{(1 - v^2)^{1/2}}, \qquad k_t = k_t' \frac{(1 + v\cos\alpha')}{(1 - v^2)^{1/2}}. \tag{34}$$

The second of (34) is the usual Doppler relation, and the result

$$\cos\alpha = \frac{\cos\alpha' + v}{1 + v\cos\alpha'} \tag{35}$$

obtained by combining both of equations (34) is the usual aberration formula.

To proceed, suppose that in some way we could be assured that the material at X' is *not* in motion with respect to X. How then should we understand $z \neq 0$? An unconventional possibility would be to take the basic length unit C to be different at the two points in question. This would imply that the electron mass was a function of position, $m_e(X)$. Although this possibility may seem strange, local laboratory experiments cannot serve to demonstrate its incorrectness, since experiments tell us nothing about the local value of C itself.

More conventionally, we could argue that the assumption of flat spacetime was in error. In place of

$$ds^2 = \eta_{ik}\, dx^i\, dx^k \tag{36}$$

for the square of the distance between points x^i and $x^i + dx^i$, we could write the Riemannian form

$$ds^2 = g_{ik}\, dx^i\, dx^k, \tag{37}$$

where the functions $g_{ik}(x)$, $i, k = 1, 2, 3, 4$, are such that no transformation of coordinates will reduce (37) to (36) at all X. A transformation of coordinates can certainly be found to reduce (37) to (36) in the neighborhood of a particular X. However, the transformation required to make a similar reduction in the neighborhood of some other point, X_0, say, will in general be different. A plane electromagnetic wave with phase factor $\exp[-i\eta_{lm}k^l x^m]$ in the locally flat region at X does not reach X_0 with the same phase factor, even when we transform to locally flat space at X_0. The form of the wave is modified by the non-Euclidean nature of the geometry in the region between X and X_0.

Apart from the Doppler shift discussed above, we therefore have two ways of understanding how z in (30) might be nonzero. The possibility of C being variable appears heterodox, but we shall now investigate a close connnection between this apparently curious possibility and the seemingly more orthodox appeal to curved spacetime. This connection arises because, as we have already seen in Chapter 2, the electromagnetic theory is invariant with respect to conformal transformations (not the same as coordinate transformations), i.e., transformations which change the *geometry* from (37) to

$$ds^{*2} = g_{ik}{}^{*}\, dx^i\, dx^k, \tag{38}$$

where

$$g_{ik}{}^{*} = \Omega^2 g_{ik} = e^{2\xi} g_{ik}. \tag{39}$$

The meaning of this invariance is that, if F_{ik} and j^m satisfy Maxwell's equations in the unstarred geometry, then

$$F^{*}_{ik} = F_{ik}, \qquad J^{*m} = \Omega^{-4} J^m, \tag{40}$$

satisfy Maxwell's equations in the starred geometry. The second of equations (40) is just that which we expect for charged particles having the same coordinate paths in both geometries. Thus,

$$J^i(X) = \left[\sum e\frac{dx^i}{ds}\right]_{\substack{\text{particles in box of}\\ \text{unit proper 3-}\\ \text{dimensional volume}\\ \text{at } X}}, \qquad J^{*i}(X) = \left[\sum e\frac{dx^i}{ds^*}\right]_{\substack{\text{similar box but}\\ \text{in starred}\\ \text{geometry}}} \tag{41}$$

Since

$$ds^* = \Omega\, ds, \tag{42}$$

and since unit proper three-dimensional volume at X contains Ω^{-3} times as many particles in the starred geometry, (41) leads immediately to $J^{*i}(X) = \Omega^{-4}J^i(X)$.

In virtue of $F_{ik}{}^* = F_{ik}$ it follows that, if a given set of coordinate paths for the charged particles produces an electromagnetic wave with phase factor

$$\exp[-i\eta_{lm}k^l x^m] \tag{43}$$

at X for the unstarred geometry, then the same set of paths in the starred geometry will also produce an electromagnetic wave at X having (43) as its phase factor. However, the metric in the vicinity of X is now scaled by $\Omega^2(X)$; so the proper frequency of the wave is changed by $\Omega(X)^{-1}$. This is just what we might have anticipated from the dimensionality C^{-1} of "frequency." Because of the conformal invariance of Maxwell's equations, proper frequency behaves with respect to conformal transformations like an inverse geometric length.

Suppose now that an observer at a point, say, X_0, is able to ascribe coordinates x^i to matter at all points X from which light is received. Can the observer hope to determine the geometry of the world, $g_{ik}(X)$, and also to discover if the physical length unit C is the same at all places? Suppose the observer imagines himself to be successful in this enterprise. Make a conformal transformation to $g_{ik}{}^*(X) = \Omega^2(X)g_{ik}(X)$, with $\Omega(X) \neq \Omega(X_0)$ for some point X. Then, provided we multiply the physical length unit at X by $\Omega(X)$ and that at X_0 by $\Omega(X_0)$, there will be no change in the results of local experiments at X_0 or at X. In particular, there will be no change in the relation of the frequencies of electromagnetic waves to the physical unit. A spectrum line emitted at X by a specific atom keeps the same relation to the physical unit at X, and it reaches X_0 with the same relation to the physical unit at X_0. Hence the conformal transformation, taken together with a scaling of the physical length unit at each point X by $\Omega(X)$, preserves all the previous relationships. Evidently then, provided we admit the possibility of a physical unit $C(X)$ varying with X, the starred geometry represents the world just as well as the unstarred geometry. The answer to our question is, therefore, no.

The situation would be different if we knew *a priori* that C must be the same at all X. This is the usual assumption of physics and cosmology. But we have no such *a priori* knowledge, nor can we expect that physical experiments will ever lead to a theory which would require such a constancy. Physical experiments, one would hope, will eventually lead to an understanding of the *mass ratios* of elementary particles, but we cannot expect to learn about C merely from the determination of dimensionless quantities, which is all that our experiments permit us to do. It seems, then, that it is inherently impossible from experiments and observations ever to determine a unique geometry for the world.

On the other hand, it may be possible to expose the usual assumptions of cosmology to test. According to conventional ideas, the tensor g_{ik} can be reduced to the Minkowski form η_{ik} over a region of spacetime with a scale of the order of the diameter of clusters of galaxies. It is also considered that C is the same at all X within such a region. Then the wavelengths of spectrum lines emitted by all objects within such a region should be the same, except for variations due to Doppler motions of the objects in question. This condition is in principle open to test. At present, the astronomical evidence is equivocal; the case of the galaxy NGC 7603 and its companion, reported by Arp (1971), is of special interest here.

§6.3. CONFORMAL INVARIANCE CONTINUED (LECTURE 15)

The dilemma we have now reached has occurred before. Newtonian theory sought to determine the physical laws with respect to a unique direction of the time-axis—it was considered that the laws should be invariant for spatial rotations, but not for general coordinate transformations. This enterprise proved to be impossible, and the impossibility became translated into a virtue, with the eventual understanding that physical laws must be invariant under general coordinate transformations. We can proceed in the same way here, by requiring the physical laws to be also invariant under conformal transformations of the geometry. All such geometries then become physically equivalent. If for convenience we choose a particular geometry (in order to investigate some problem), our choice has no more absolute significance than has the choice of a particular coordinate system.

Since particle dynamics as now formulated, whether classically or in terms of quantum mechanics, is *not* invariant under conformal transformations, our present enterprise must involve a new approach to physics. The lack of conformal invariance shows classically in the form (13) for the action S. Since da changes to $da^* = \Omega\, da$ under a conformal transformation from g_{ik} to $g_{ik}{}^* = \Omega^2 g_{ik}$, it is clear that S is not invariant under such a transformation unless m is changed to $m^* = \Omega^{-1}m$. This change contradicts the usual idea that particle masses are fixed and inviolable.

The same situation occurs in quantum mechanics. The Dirac equation in curved space is conformally invariant for

$$m^* = \Omega^{-1}m, \qquad \psi^* = \Omega^{-3/2}\psi, \tag{44}$$

but not for fixed m. It will be noticed that the second relation in (44) requires that the probability of finding the particle in question in a three-dimensional box of unit proper volume must change by Ω^{-3} in going from the unstarred to the starred geometry. This is just what we would expect from the changed relation between coordinate volume and proper volume. Once again the meaning of (44) is that if ψ, m, satisfy Dirac's equation in the unstarred geometry, then ψ^*, m^*, satisfy the Dirac equation in the starred geometry. This statement is proved in Appendix 4 (p. 254).

§6.4. INERTIA

The transformation $m^* = \Omega^{-1}m$ demands a physical interpretation of mass. To obtain such an interpretation, we now take a Machian view, according to which the mass of a particle arises from interactions with other particles, and possibly also from self-interaction. Although we use the term "Machian," our position now is much stronger than the rather vague philosophical arguments originally given by Mach.

Let each particle b give rise to what we might term a "mass field." Denote this field at a general point X by $m^{(b)}(X)$. At any point A on the path of particle a, we have $m^{(b)}(A)$ as the contribution of particle b to the mass of particle a at the point A. Summing for all b gives

$$m(A) = \sum_b m^{(b)}(A). \tag{45}$$

The non-electromagnetic part of the action, S_{mat}, for a set of particles a, $b, \ldots,$ can be written in the form

$$S_{\text{mat}} = -\frac{1}{2}\sum_a \int m(A)\, da = -\frac{1}{2}\sum_a \sum_b \int m^{(b)}(A)\, da. \tag{46}$$

In order that (46) be symmetric for any particle pair a, b, we must have $m^{(b)}(A)$ of the form

$$m^{(b)}(A) = \int P(A, B)\, db, \tag{47}$$

where $P(A, B)$ is a biscalar symmetric with respect to A, B. Then we can write (46) in the symmetric form

$$S_{\text{mat}} = -\frac{1}{2} \sum_a \sum_b \int \int P(A, B)\, da\, db, \tag{48}$$

from which it is clear that S_{mat} is conformally invariant provided

$$P^*(A, B) = \Omega(A)^{-1} \Omega(B)^{-1} P(A, B). \tag{49}$$

The factor $\frac{1}{2}$ has been introduced into (46) in analogy to (29) of Chapter 5, because the pair a, b, appears twice in the double sum.

Also, in analogy to the electromagnetic case, we permit $a = b$ in (46). It was found necessary in Chapter 4 to prevent $A \equiv B$ in such a case, in order to avoid infinities. We could adopt the same point of view here, but doing so may not be necessary. The theory here will include gravitation, which was omitted in Chapter 4, except that a cutoff, defined gravitationally, was used there in order to limit the approach of B to A. However, the theory here will not be fully quantum; so this issue remains uncertain.

The function $P(X, B)$ propagates the mass field of particle b, and we expect it to satisfy a wave equation of the usual kind. It turns out that for (49) to hold, $P(X, B)$ must be of the form (constant) $\cdot\, \widetilde{G}(X, B)$, $\widetilde{G}(X, B)$ being a function with the properties

$$\Box_X \widetilde{G}(X, B) + \frac{1}{6} R(X) \widetilde{G}(X, B) = [-g(B)]^{-1/2}\, \delta_4(X, B), \tag{50}$$

$$\widetilde{G}(X, B) = \widetilde{G}(B, X), \tag{51}$$

where $\delta_4(X, B)$ is the four-dimensional delta function representing a source of unit strength at B, and $R(X)$ is the Riemannian scalar curvature at X. In order that masses shall satisfy the usual convention of being positive, we take the constant of proportionality between $P(X, B)$ and $\widetilde{G}(X, B)$ to be positive,

$$P(X, B) = \lambda^2 \widetilde{G}(X, B), \qquad \lambda \text{ real.} \tag{52}$$

In flat space $R(X) = 0$, and the solution for $\widetilde{G}(X, B)$ is just

$$\widetilde{G}(X, B) = \frac{1}{4\pi} \delta(s^2{}_{XB}). \tag{53}$$

We now have

$$S_{\text{mat}} = \frac{-\lambda^2}{2} \sum_a \sum_b \int\int \widetilde{G}(A, B)\, da\, db. \tag{54}$$

From (49), (51), (52), we require

$$\widetilde{G}^*(X, B) = \widetilde{G}^*(B, X) = \Omega(X)^{-1}\Omega(B)^{-1}\widetilde{G}(X, B). \tag{55}$$

The meaning of (55) is that if $\widetilde{G}(X, B) = \widetilde{G}(B, X)$ satisfies (50), then $\widetilde{G}^*(X, B)$ given by (55) satisfies the corresponding wave equation in the starred geometry; we end this section by proving this result.

First, we recall (77) of Chapter 2,

$$R^* = \Omega^{-2}[R + 6\Omega^{-1}\Box\Omega], \tag{56}$$

so that

$$\frac{1}{6} R^*(X)\widetilde{G}^*(X, B)$$

$$= \frac{1}{6} \Omega^{-1}(B)\Omega^{-3}(X)\widetilde{G}(X, B)[R(X) + 6\Omega^{-1}(X)\Box_X\Omega(X)]. \tag{57}$$

Using

$$g^{*ik} = \Omega^{-2}g^{ik}, \qquad \sqrt{-g^*} = \Omega^4\sqrt{-g}, \tag{58}$$

we have

$$\Box_X{}^*\widetilde{G}^*(X, B) = \frac{1}{\sqrt{-g^*(X)}} \frac{\partial}{\partial x^i}\left[\sqrt{-g^*(X)}\, g^{*ik}(X) \frac{\partial \widetilde{G}^*(X, B)}{\partial x^k}\right]$$

$$= \frac{\Omega^{-1}(B)\Omega^{-4}(X)}{\sqrt{-g(X)}} \frac{\partial}{\partial x^i}\left[\Omega^2(X)\sqrt{-g(X)}\, g^{ik}(X) \frac{\partial}{\partial x^k}\left(\frac{\widetilde{G}(X, B)}{\Omega(X)}\right)\right]$$

$$= \Omega^{-1}(B)\Omega^{-4}(X)[\Omega(X)\Box_X\widetilde{G}(X, B) - \widetilde{G}(X, B)\Box_X\Omega(X)]. \tag{59}$$

Adding (57) and (59), the second terms on the righthand sides cancel, giving

$$\Box_X{}^*\widetilde{G}^*(X, B) + \frac{1}{6}R^*(X)\widetilde{G}^*(X, B)$$

$$= \Omega^{-1}(B)\Omega^{-3}(X)\left[\Box_X\widetilde{G}(X, B) + \frac{1}{6}R(X)\widetilde{G}(X, B)\right]$$

$$= \frac{\Omega^{-1}(B)\Omega^{-3}(X)}{\sqrt{-g(B)}}\,\delta_4(X, B) = \frac{\delta_4(X, B)}{\sqrt{-g^*(B)}}. \tag{60}$$

This completes the proof.

§6.5. GRAVITATION (LECTURE 16)

In the Einstein theory one starts with an action formula

$$S = \frac{1}{16\pi G}\int R\sqrt{-g}\,d^4x - \sum_a m_a \int da \tag{61}$$

in which the mass m_a of particle a belongs autonomously to the particle, and is a fixed quantity. The constant G is also introduced *ad hoc,* and is taken to be positive in order that gravity will turn out to be attractive. A variation

$$g_{ik} \to g_{ik} + \delta g_{ik} \tag{62}$$

is then considered within a finite 4-volume V, with δg_{ik} satisfying the usual conditions of continuity and differentiability. We have

$$\delta S = \frac{1}{16\pi G}\int_V \delta(R\sqrt{-g})\,d^4x - \sum_a m_a\,\delta\int da$$

$$= \frac{1}{16\pi G}\int_V \delta(g^{ik}R_{ik}\sqrt{-g})\,d^4x - \sum_a m_a\,\delta\int da. \tag{63}$$

As in (32) of Chapter 5, the energy-momentum tensor $T_{(\mathrm{mat})}{}^{ik}$ of the particles is defined by

$$-\sum_a m_a\,\delta\int da = -\frac{1}{2}\int T_{(\mathrm{mat})}{}^{ik}\,\delta g_{ik}\sqrt{-g}\,d^4x$$

$$= \frac{1}{2}\int T_{(\mathrm{mat})ik}\,\delta g^{ik}\sqrt{-g}\,d^4x, \tag{64}$$

the last step of (64) following from $\delta g_{ik} = -g_{ip}g_{kq}\,\delta g^{pq}$. It can be shown without undue difficulty that

$$T_{(\text{mat})}{}^{ik}(X) = \sum_a m_a \int \frac{\delta_4(X, A)}{[-g(A)]^{1/2}} \frac{da^i}{da}\frac{da^k}{da}\,da. \tag{65}$$

Noting that $\delta\sqrt{-g} = -\tfrac{1}{2}g_{ik}\,\delta g^{ik}\sqrt{-g}$, we get

$$\delta S = \frac{1}{16\pi G}\int_V \left(R_{ik} - \frac{1}{2}g_{ik}R\right)\delta g^{ik}\sqrt{-g}\,d^4x$$
$$+\frac{1}{2}\int_V T_{(\text{mat})ik}\,\delta g^{ik}\sqrt{-g}\,d^4x + \frac{1}{16\pi G}\int_V g^{ik}\,\delta R_{ik}\sqrt{-g}\,d^4x. \tag{66}$$

A straightforward calculation then shows that the last term on the righthand side of (66) is zero, so that

$$\delta S = \frac{1}{16\pi G}\int_V \left(R_{ik} - \frac{1}{2}g_{ik}R + 8\pi G\,T_{(\text{mat})ik}\right)\delta g^{ik}\sqrt{-g}\,d^4x. \tag{67}$$

Setting

$$\delta S = 0 \tag{68}$$

for δg^{ik} satisfying the conditions of continuity and differentiability but otherwise arbitrary, and for V arbitrary, then requires

$$R_{ik} - \frac{1}{2}g_{ik}R + 8\pi G\,T_{(\text{mat})ik} = 0, \tag{69}$$

which are the Einstein equations.

In the direct-particle theory, in place of (61) we have the simpler expression

$$S = -\frac{1}{2}\sum_a \int m(A)\,da = -\frac{\lambda^2}{2}\sum_a\sum_b \int\int \tilde{G}(A, B)\,da\,db. \tag{70}$$

Subject to (68), can this be made to lead to the Einstein equations? Since there is no *ad hoc* constant G in (70), there cannot be a precise equivalence to the Einstein theory. The coupling constant λ^2—which is so far arbitrary—does not take the place of G, since λ^2 will simply cancel from (68). It follows that, if we can obtain equations of the form of (69) from (70), the quantity G must turn out to be calculable. It can be seen from (61) that G must have dimensionality L^2. It is therefore not on the same footing as the dimensionless fine-structure constant e^2, and must change as Ω^2 in a conformal transformation—if indeed our physics is to be conformally invariant.

Although (70) is simpler in its statement than (61), the calculation of δS for (70) is more complicated. The equation satisfied by $\widetilde{G}(X, B)$ is

$$\frac{\partial}{\partial x^{i_x}}\left[\sqrt{-g(X)}g^{i_x k_x}\frac{\partial \widetilde{G}(X, B)}{\partial x^{k_x}}\right] + \frac{1}{6}R(X)\sqrt{-g(X)}\widetilde{G}(X, B) = \delta_4(X, B). \quad (71)$$

The variation of (71) can be written

$$\frac{\partial}{\partial x^i}\left[\sqrt{-g}\, g^{ik}\frac{\partial}{\partial x^k}\delta\widetilde{G}\right] + \frac{1}{6}R\sqrt{-g}\,\delta\widetilde{G} = -\frac{\partial}{\partial x^i}\left[\delta(\sqrt{-g}\, g^{ik})\frac{\partial \widetilde{G}}{\partial x^k}\right] - \frac{1}{6}\delta(R\sqrt{-g})\widetilde{G}. \quad (72)$$

Just as in §5.2, we deal with the source term on the righthand side of (72) by splitting $\widetilde{G}$ into retarded and advanced parts,

$$\widetilde{G}(X, B) = \frac{1}{2}[G^{(\text{ret})}(X, B) + G^{(\text{adv})}(X, B)]. \quad (73)$$

The retarded part gives the following contribution to $\delta\widetilde{G}(A, B)$:

$$-\frac{1}{2}\int_V G^{(\text{ret})}(A, X)\frac{\partial}{\partial x^i}\left[\delta(\sqrt{-g}\, g^{ik})\frac{\partial G^{(\text{ret})}(X, B)}{\partial x^k}\right]d^4x - \frac{1}{12}\int_V \delta(R\sqrt{-g})G^{(\text{ret})}(A, X)G^{(\text{ret})}(X, B)\, d^4x. \quad (74)$$

From the divergence theorem, and noting that

$$G^{(\text{ret})}(A, X) = G^{(\text{adv})}(X, A), \quad (75)$$

(74) can be written more symmetrically as

$$\frac{1}{2}\int_V \delta(\sqrt{-g}\, g^{ik})\frac{\partial}{\partial x^i}[G^{(\text{adv})}(X, A)]\frac{\partial}{\partial x^k}[G^{(\text{ret})}(X, B)]\, d^4x - \frac{1}{12}\int_V \delta(R\sqrt{-g})G^{(\text{adv})}(X, A)G^{(\text{ret})}(X, B)\, d^4x. \quad (76)$$

The advanced part of (73) gives a similar contribution to $\delta\widetilde{G}(A, B)$, but with the advanced and retarded forms of G in (76) inverted:

$$\frac{1}{2}\int_V \delta(\sqrt{-g}\, g^{ik})\frac{\partial}{\partial x^i}[G^{(\text{ret})}(X, A)]\frac{\partial}{\partial x^k}[G^{(\text{adv})}(X, B)]\, d^4x$$
$$-\frac{1}{12}\int_V \delta(R\sqrt{-g})G^{(\text{ret})}(X, A)G^{(\text{adv})}(X, B)\, d^4x. \quad (77)$$

Then $\delta\widetilde{G}(A, B)$ is the sum of (76) and (77).

The next step is to obtain

$$\lambda^2 \sum_a \sum_b \int\int \delta\widetilde{G}(A, B)\, da\, db. \quad (78)$$

Defining a "mass field"

$$M(X) = \frac{1}{2}[M^{(\text{ret})}(X) + M^{(\text{adv})}(X)], \quad (79)$$

where

$$M^{(\text{ret})}(X) = \lambda \sum_a \int G^{(\text{ret})}(X, A)\, da, \quad (80)$$

$$M^{(\text{adv})}(X) = \lambda \sum_a \int G^{(\text{adv})}(X, A)\, da, \quad (81)$$

(78) simplifies to

$$\int_V \left[\delta(\sqrt{-g}\, g^{ik})M_i^{(\text{ret})}M_k^{(\text{adv})} - \frac{1}{6}\delta(R\sqrt{-g})M^{(\text{ret})}M^{(\text{adv})}\right] d^4x, \quad (82)$$

where

$$M_i^{(\text{ret})} = \frac{\partial M^{(\text{ret})}}{\partial x^i}, \qquad M_k^{(\text{adv})} = \frac{\partial M^{(\text{adv})}}{\partial x^k}. \quad (83)$$

Remembering the factor λ^2 in (52), the mass $m(A)$ of particle a at point A is $\lambda M(A)$, and

$$\delta S = -\frac{1}{2}\delta\Big[\sum_a \int m(A)\,da\Big]$$

$$= -\frac{1}{2}\sum_a \int m(A)\,\delta(da) - \frac{1}{2}\sum_a \int \delta m(A)\,da$$

$$= -\frac{1}{2}\int_V T_{(\text{mat})}{}^{ik}\,\delta g_{ik}\sqrt{-g}\,d^4x - \frac{1}{2}\lambda^2 \sum_a \sum_b \int\int \delta\widetilde{G}(A, B)\,da\,db$$

$$= \frac{1}{2}\int_V T_{(\text{mat})ik}\,\delta g^{ik}\sqrt{-g}\,d^4x$$

$$-\frac{1}{2}\int_V \Big[\delta(\sqrt{-g}\,g^{ik})M_i{}^{(\text{ret})}M_k{}^{(\text{adv})}$$

$$-\frac{1}{6}\delta(R\sqrt{-g})M^{(\text{ret})}M^{(\text{adv})}\Big]\,d^4x. \tag{84}$$

The energy momentum tensor

$$T_{(\text{mat})}{}^{ik}(X) = \sum_a \int m(A)\frac{\delta_4(X, A)}{[-g(A)]^{1/2}}\frac{da^i}{da}\frac{da^k}{da}\,da, \tag{85}$$

is obtained in the same way as in the Einstein theory. The mass $m(A)$ can depend now on point A, however, and must therefore be placed within the line integral.

At this stage we might pause to note that we are here considering all particles to have the same "coupling constant" λ to the mass field. It is easy to modify the above discussion to deal with a case in which particle a has coupling constant λ_a, particle b has coupling constant λ_b, and so on. We change the definitions (80), (81), to

$$M^{(\text{ret})}(X) = \sum_a \lambda_a \int G^{(\text{ret})}(X, A)\,da, \tag{86}$$

$$M^{(\text{adv})}(X) = \sum_a \lambda_a \int G^{(\text{adv})}(X, A)\,da, \tag{87}$$

and define the mass of particle a at point A by

$$m_a(A) = \lambda_a M(A). \tag{88}$$

Then

$$\delta S = -\frac{1}{2}\delta\left[\sum_a \int m_a(A)\, da\right] \tag{89}$$

continues to be given by (84), provided $m(A)$ in (85) is replaced by $m_a(A)$.

To obtain the gravitational equations, we have to express the second term on the righthand side of (84) in the form

$$\frac{1}{2}\int_V Q_{ik}\, \delta g^{ik} \sqrt{-g}\, d^4x. \tag{90}$$

The gravitational equations, given again by putting $\delta S = 0$ for arbitrary δg^{ik}, are simply

$$Q_{ik} + T_{(\mathrm{mat})ik} = 0. \tag{91}$$

The term involving $M_i^{(\mathrm{ret})} M_k^{(\mathrm{adv})}$ is easily seen to give the contribution

$$-\frac{1}{2}[M_i^{(\mathrm{ret})} M_k^{(\mathrm{adv})} + M_i^{(\mathrm{adv})} M_k^{(\mathrm{ret})} - g_{ik} g^{pq} M_p^{(\mathrm{adv})} M_q^{(\mathrm{ret})}]. \tag{92}$$

Writing

$$\begin{aligned} \delta(R\sqrt{-g}) &= \delta(R_{ik} g^{ik} \sqrt{-g}) \\ &= \left(R_{ik} - \frac{1}{2} g_{ik} R\right) \delta g^{ik} \sqrt{-g} + g^{ik}\, \delta R_{ik} \sqrt{-g}, \end{aligned} \tag{93}$$

we have a further contribution

$$\frac{1}{6}\left(R_{ik} - \frac{1}{2} g_{ik} R\right) M^{(\mathrm{ret})} M^{(\mathrm{adv})}, \tag{94}$$

the remaining term of (84) being

$$\frac{1}{12}\int_V \delta R_{ik} g^{ik}\, M^{(\mathrm{ret})} M^{(\mathrm{adv})} \sqrt{-g}\, d^4x. \tag{95}$$

If we express (95) in the form

$$\frac{1}{2}\int \Theta_{ik}\, \delta g^{ik} \sqrt{-g}\, d^4x, \tag{96}$$

the gravitational equations become

$$T_{(\mathrm{mat})ik} + \Theta_{ik} + \frac{1}{6}\left(R_{ik} - \frac{1}{2} g_{ik} R\right) M^{(\mathrm{ret})} M^{(\mathrm{adv})}$$
$$- \frac{1}{2}[M_i^{(\mathrm{ret})} M_k^{(\mathrm{adv})} + M_k^{(\mathrm{ret})} M_i^{(\mathrm{adv})} - g_{ik} g^{pq} M_q^{(\mathrm{ret})} M_p^{(\mathrm{adv})}] = 0. \quad (97)$$

As in the Einstein theory, we begin the reduction of (95) with

$$g^{ik}\, \delta R_{ik} = -\frac{1}{\sqrt{-g}} \frac{\partial}{\partial x^l} [\sqrt{-g}\, w^l], \quad (98)$$

where

$$w^l = g^{ik}\, \delta \Gamma^l{}_{ik} - g^{il}\, \delta \Gamma^k{}_{ik}. \quad (99)$$

Using the divergence theorem we can therefore write (95) as

$$\frac{1}{12} \int_V w^l \frac{\partial}{\partial x^l} [M^{(\mathrm{ret})} M^{(\mathrm{adv})}] \sqrt{-g}\, d^4x. \quad (100)$$

And using the identities

$$\Gamma^k{}_{ik} = \frac{1}{\sqrt{-g}} \frac{\partial}{\partial x^i} (\sqrt{-g}), \qquad g^{ik} \Gamma^l{}_{ik} = -\frac{1}{\sqrt{-g}} \frac{\partial}{\partial x^k} (\sqrt{-g}\, g^{lk}), \quad (101)$$

we see that

$$w^l = -\delta g^{ik} \left[\Gamma^l{}_{ik} + \frac{1}{2} g_{ik} g^{pl} \Gamma^q{}_{pq} - \frac{1}{2} g_{ik} g^{pq} \Gamma^l{}_{pq}\right]$$
$$- \frac{1}{\sqrt{-g}} \frac{\partial}{\partial x^i} [\delta(g^{il} \sqrt{-g})] - \frac{1}{\sqrt{-g}} g^{il} \frac{\partial}{\partial x^i} (\delta \sqrt{-g}). \quad (102)$$

Substituting (102) in (100), and making straightforward reductions in which the divergence theorem is again used, we obtain

$$-\frac{1}{12} \int_V \{g_{ik}\, \Box [M^{(\mathrm{ret})} M^{(\mathrm{adv})}] - [M^{(\mathrm{ret})} M^{(\mathrm{adv})}]_{;ik}\}\, \delta g^{ik} \sqrt{-g}\, d^4x. \quad (103)$$

The gravitational equations therefore take the form

$$\frac{1}{6}\left(R_{ik}-\frac{1}{2}g_{ik}R\right)[M^{(\text{ret})}M^{(\text{adv})}]+T_{(\text{mat})ik}$$
$$-\frac{1}{2}[M_i^{(\text{ret})}M_k^{(\text{adv})}+M_k^{(\text{ret})}M_i^{(\text{adv})}-g_{ik}g^{pq}M_p^{(\text{ret})}M_q^{(\text{adv})}]$$
$$-\frac{1}{6}\{g_{ik}\,\Box[M^{(\text{ret})}M^{(\text{adv})}]-[M^{(\text{ret})}M^{(\text{adv})}]_{;ik}\}=0. \qquad (104)$$

It is of interest to consider the contracted form of (104),

$$-\frac{1}{6}R[M^{(\text{ret})}M^{(\text{adv})}]+T_{(\text{mat})}+g^{pq}M_p^{(\text{ret})}M_q^{(\text{adv})}$$
$$-\frac{1}{2}\Box[M^{(\text{ret})}M^{(\text{adv})}]=0. \qquad (105)$$

Writing

$$m(A)=\lambda M(A)=\frac{1}{2}\lambda[M^{(\text{ret})}(A)+M^{(\text{adv})}(A)] \qquad (106)$$

in (85),

$$T_{(\text{mat})}=\frac{1}{2}\lambda\sum_a\int[M^{(\text{ret})}(A)+M^{(\text{adv})}(A)]\frac{\delta_4(X,A)}{[-g(A)]^{1/2}}\,da. \qquad (107)$$

Furthermore

$$\Box[M^{(\text{ret})}M^{(\text{adv})}]$$
$$=M^{(\text{ret})}\Box M^{(\text{adv})}+M^{(\text{adv})}\Box M^{(\text{ret})}+2g^{pq}M_p^{(\text{ret})}M_q^{(\text{adv})}=0, \qquad (108)$$

so that the term in $g^{pq}M_p^{(\text{ret})}M_q^{(\text{adv})}$ cancels from the lefthand side of (105). Moreover, the definitions (80), (81), of $M^{(\text{ret})}$, $M^{(\text{adv})}$, taken with the fact that $\tilde{G}^{(\text{ret})}$, $\tilde{G}^{(\text{adv})}$, are the retarded and advanced solutions of (71), lead to

$$\Box M^{(\text{ret})}(X)+\frac{1}{6}R(X)M^{(\text{ret})}(X)=\lambda\sum_a\int\frac{\delta_4(X,A)}{[-g(A)]^{1/2}}\,da, \qquad (109)$$

$$\Box M^{(\text{adv})}(X)+\frac{1}{6}R(X)M^{(\text{adv})}(X)=\lambda\sum_a\int\frac{\delta_4(X,A)}{[-g(A)]^{1/2}}\,da. \qquad (110)$$

Hence the lefthand side of (105) vanishes identically. What can this result mean?

The aim of the Einstein theory is to determine both the geometry, expressed by the symmetric metric tensor $g_{ik}(x)$, and the trajectories of the particles. At first sight there might seem to be ten independent components of g_{ik} at each point X required to determine the geometry. Yet the coordinate transformation $x'^i = x'^i(x)$ changes the g_{ik} to g_{ik}', where

$$g_{ik}' = \frac{\partial x^l}{\partial x'^i}\frac{\partial x^m}{\partial x'^k} g_{lm}, \tag{111}$$

and the tensor g_{ik}' determines the geometry just as well as g_{ik}. Hence not all ten components of g_{ik} are needed to determine the geometry—the ten components contain a redundancy which expresses our freedom to make coordinate transformations. In fact, only six independent functions are needed at each point X in order to specify the geometry. This is a question that we shall consider further later on.

Yet the Einstein equations (69) at first sight seem to have ten independent components. These do not overconstrain the determination of g_{ik}, however, because of the identities

$$\left(R^{ik} - \frac{1}{2} g^{ik} R\right)_{;k} \equiv 0, \tag{112}$$

which require

$$T_{(\mathrm{mat});k}{}^{ik} = 0. \tag{113}$$

Equation (113) has four components that determine the trajectories of the particles. Thus the ten Einstein equations can be thought of as comprising six for the geometry and four for the particle trajectories.

In the theory here our aim is not to determine a unique geometry. Because of the conformal invariance of the theory, we can determine the geometry only to within a conformal transformation. This reduces the number of functions needed at each point X from six to five. The ten components of (104) contain four conditions to determine the particle trajectories, five to determine the family of conformally equivalent geometries, and one identity. The identity appeared when we took the contracted form of (104). Thus we have explained the result we reached with (110).

We come now to the critical question: Can we choose a particular geometry belonging to our family of conformally equivalent geometries in which the gravitational equations (104) become the same as the Einstein equations?

Provided we introduce a "response condition" of exactly the same form as that used in §5.2 for the electromagnetic energy-momentum, the answer to this question is yes. Let us follow the procedure of §5.2, simply replacing $\Sigma_a F^{(a)(\text{ret})}$, $\Sigma_a F^{(a)(\text{adv})}$, by $M^{(\text{ret})}$, $M^{(\text{adv})}$. Thus consider the slab of spacetime $0 \leq t \leq T$. For any point X in the slab, we have

$$M^{(\text{ret})} = M^{(\text{ret})}{}_{0 \leq t \leq T} + M^{(\text{ret})}{}_{t<0},$$
$$M^{(\text{adv})} = M^{(\text{adv})}{}_{0 \leq t \leq T} + M^{(\text{adv})}{}_{t>T}. \tag{114}$$

As in §5.2, we suppose for simplicity that the retarded fields "start" at $t = 0$. This is expressed by putting $M^{(\text{ret})}{}_{t<0} = 0$. For $M^{(\text{adv})}{}_{t>T}$ we use a response condition similar to (52),

$$M^{(\text{adv})}{}_{t>T} = M^{(\text{ret})}{}_{0 \leq t \leq T} - M^{(\text{adv})}{}_{0 \leq t \leq T}. \tag{115}$$

Then

$$M^{(\text{adv})} = M^{(\text{ret})} = M^{(\text{ret})}{}_{0 \leq t \leq T}, \tag{116}$$

and we can write (104) in the form

$$K\left(R_{ik} - \frac{1}{2} g_{ik} R\right) = -3[T_{(\text{mat})ik} + \Phi_{ik}] + (g_{ik} \Box K - K_{;ik}), \tag{117}$$

where

$$K = \frac{1}{2} [M^{(\text{ret})}]^2, \tag{118}$$

$$\Phi_{ik} = -M_i^{(\text{ret})} M_k^{(\text{ret})} + \frac{1}{2} g_{ik} g^{pq} M_p^{(\text{ret})} M_q^{(\text{ret})}, \tag{119}$$

and where $T_{(\text{mat})ik}$ can be written in the form

$$T_{(\text{mat})ik} = \lambda \sum_a \int M^{(\text{ret})}(A) \frac{\delta_4(X, A)}{[-g(A)]^{1/2}} \frac{da_i}{da} \frac{da_k}{da}\, da, \tag{120}$$

so that only $M^{(\text{ret})}$ appears in (117). Now because the theory is conformally invariant, we can choose a particular conformal frame in which $M^{(\text{ret})}(X)$ is independent of X, since the mass field $M^{(\text{ret})}$ transforms as Ω^{-1} in a conformal transformation. If $M^{(\text{ret})}$ is not constant to begin with, we can evidently make $M^{*(\text{ret})}$ constant by a conformal transformation with $\Omega \propto M^{(\text{ret})}$. Then (117) simplifies to

$$K\left(R_{ik} - \frac{1}{2} g_{ik} R\right) = -3T_{(\mathrm{mat})ik}, \tag{121}$$

which has the form of the Einstein equations. Moreover, the particle masses λM^{ret} are now constant, as in Einstein's theory. Indeed, the only difference from Einstein's theory is that in this theory we have an explicit determination of the gravitational constant. Thus defining

$$8\pi G = 3/K, \tag{122}$$

the equations (121) become the same as (69). Since $M^{(\mathrm{ret})}$ is in principle calculable for any given cosmological model, K, and hence G, is calculable. We shall consider such calculations for various models in the next chapter. Here we simply notice that the definition (118) requires K, and hence G, to be positive. It follows that we have *deduced* that gravitation must be attractive, whereas in the Einstein theory the assumption $G > 0$ was introduced *ad hoc*.

The appearance of λ in (120) does *not* mean that the value of this coupling constant affects a purely gravitational theory. Since $M^{(\mathrm{ret})}$ is also proportional to λ, every term in (117) has a λ^2 dependence, so that λ^2 could be canceled throughout (117). If, however, we were also to include an electromagnetic field, with (117) changing to

$$K\left(R_{ik} - \frac{1}{2} g_{ik} R\right) = -3[T_{(\mathrm{mat})ik} + T_{(\mathrm{em})ik} + \Phi_{ik}] + (g_{ik} \Box K - K_{;ik}), \tag{123}$$

we would have λ^2 in all terms except $T_{(\mathrm{em})ik}$. The value given to λ^2 then affects the relative importance of $T_{(\mathrm{em})ik}$ to the gravitational field: the smaller λ^2 is, the more important the electromagnetic field becomes. Also in a reduction to the Einstein form

$$K\left(R_{ik} - \frac{1}{2} g_{ik} R\right) = -3[T_{(\mathrm{mat})ik} + T_{(\mathrm{em})ik}], \tag{124}$$

with K now constant, we have

$$[T_{(\mathrm{mat})}{}^{ik} + T_{(\mathrm{em})}{}^{ik}]_{;k} = 0, \tag{125}$$

and the value attributed to λ^2 affects the orbits of charged particles.

The discussion here has shown that the concept of a "response of the universe" has application outside electrodynamics. Later we shall find a further example in which a response condition plays a crucial role. All these response conditions are structurally similar to each other, just as (115) is similar to (52) of Chapter 5. Both conditions are classical, however, and just as the quantum-mechanical response of the universe, given by (124) of Chapter 4, contains subtleties not present in (52) of Chapter 5, so we should expect a quantum form of (115) to open out a range of phenomena analogous in richness to quantum electrodynamics. In this book we shall not attempt anything as elaborate as a full quantum treatment of inertia and gravitation, although we shall take up some questions preliminary to such a treatment in Chapter 8. Before coming to these questions, we proceed in Chapter 7 to apply the classical development we have reached to well-known cosmological models.

Equations of Motion

The usual procedure for obtaining the equations of motion of the particles is to go back to the action (70) and to vary the particle paths individually, requiring $\delta S = 0$ for such variations. However, this procedure obscures the fact that the equations of motion are contained within the gravitational equations. Here we shall give a direct derivation from the gravitational equations, using first the conformal frame in which the particle masses m_a, $m_b, \ldots,$ are constant. In this frame, and in a purely gravitational field,

$$T_{(\mathrm{mat});k}{}^{ik} = 0 \tag{126}$$

is the required equation. Here

$$T_{(\mathrm{mat})}{}^{ik}(X) = \sum_a m_a \int \frac{\delta_4(X, A)}{\sqrt{-g(A)}} \frac{da^i}{da} \frac{da^k}{da} \, da, \tag{127}$$

so that

$$\begin{aligned}
\frac{\partial T^{ik}}{\partial x^k} &= \sum_a m_a \int \frac{\partial}{\partial x^k} [\delta_4(X, A)] \frac{1}{\sqrt{g(A)}} \frac{da^i}{da} \frac{da^k}{da} \, da \\
&= -\sum_a m_a \int \frac{\partial}{\partial a^k} [\delta_4(X, A)] \frac{1}{\sqrt{-g(A)}} \frac{da^i}{da} \frac{da^k}{da} \, da \\
&= -\sum_a m_a \int \frac{d}{da} [\delta_4(X, A)] \frac{1}{\sqrt{-g(A)}} \frac{da^i}{da} \, da \\
&= \sum_a m_a \int \delta_4(X, A) \frac{d}{da} \left(\frac{1}{\sqrt{-g(A)}} \frac{da^i}{da} \right) da,
\end{aligned} \tag{128}$$

since the particle paths are not considered to start or to end at the point X. Thus

$$T^{ik}{}_{;k} = \frac{\partial T^{ik}}{\partial x^k} + \Gamma^i{}_{kl}\, T^{lk} + \Gamma^k{}_{kl}\, T^{il}$$

$$= \sum_a m_a \int \frac{\delta_4(X, A)}{\sqrt{-g(A)}} \left[\sqrt{-g(A)}\, \frac{d}{da}\left(\frac{1}{\sqrt{-g(A)}} \frac{da^i}{da}\right) + \Gamma^i{}_{kl} \frac{da^k}{da} \frac{da^l}{da} + \Gamma^k{}_{kl} \frac{da^i}{da} \frac{da^l}{da} \right] da, \quad (129)$$

and (126) requires

$$\sqrt{-g}\, \frac{d}{da}\left(\frac{1}{\sqrt{-g}} \frac{da^i}{da}\right) + \Gamma^i{}_{kl} \frac{da^k}{da} \frac{da^l}{da} + \Gamma^k{}_{lk} \frac{da^i}{da} \frac{da^l}{da} = 0. \quad (130)$$

Using

$$\Gamma^k{}_{kl} = \frac{1}{\sqrt{-g}} \frac{\partial}{\partial x^l} \sqrt{-g}, \quad (131)$$

(130) simplifies to the usual geodesic equations

$$\frac{d^2 a^i}{da^2} + \Gamma^i{}_{kl} \frac{da^k}{da} \frac{da^l}{da} = 0. \quad (132)$$

To obtain the equations of motion in a general conformal frame, consider the effect on (132) of the transformation

$$g_{ik}{}^* = \Omega^2 g_{ik}. \quad (133)$$

Since

$$da^* = \Omega\, da, \quad (134)$$

and

$$\Gamma^i{}_{kl} = \Gamma^{*i}{}_{kl} + \left(g^{*in} g_{kl}{}^* \frac{\partial \ln \Omega}{\partial x^n} - g^{*i}{}_k \frac{\partial \ln \Omega}{\partial x^l} - g^{*i}{}_l \frac{\partial \ln \Omega}{\partial x^k}\right), \quad (135)$$

we have

$$\frac{1}{\Omega} \frac{d}{da^*}\left(\frac{\Omega\, da^i}{da^*}\right) + \Gamma^{*i}{}_{kl} \frac{da^k}{da^*} \frac{da^l}{da^*} + \left(g^{*in} g_{kl}{}^* \frac{\partial \ln \Omega}{\partial x^n} - g^{*i}{}_k \frac{\partial \ln \Omega}{\partial x^l} - g^{*i}{}_l \frac{\partial \ln \Omega}{\partial x^k}\right) \frac{da^k}{da^*} \frac{da^l}{da^*} = 0 \quad (136)$$

for the equations of motion in the starred geometry. The transformation function Ω can be replaced by m_a/m_a^*, with m_a constant, so that

$$m_a^* \frac{d}{da^*}\left(\frac{1}{m_a^*}\frac{da^i}{da^*}\right) + \Gamma^{*i}{}_{kl}\frac{da^k}{da^*}\frac{da^l}{da^*}$$
$$+\frac{1}{m_a^*}\left(g^{*i}{}_k\frac{\partial m_a^*}{\partial x^l} + g^{*i}{}_l\frac{\partial m_a^*}{\partial x^k} - g^{*in}g_{kl}^*\frac{\partial m_a^*}{\partial x^n}\right)\frac{da^k}{da^*}\frac{da^l}{da^*} = 0. \quad (137)$$

This simplifies to

$$\frac{d}{da^*}\left(m_a^*\frac{da^i}{da^*}\right) + m_a^*\Gamma^{*i}{}_{kl}\frac{da^k}{da^*}\frac{da^l}{da^*} - g^{ik}\frac{\partial m_a^*}{\partial x^k} = 0. \quad (138)$$

Thus in a general conformal frame the equations of motion are not geodesic.

It is possible to obtain (138) directly by dividing (117) through by K, and then taking the divergence. The reduction is somewhat awkward, however, it being necessary to use the wave equation for the masses, the relation (41) of Chapter 2, and also (117) itself.

7

COSMOLOGY

§7.1. THE EINSTEIN–DE SITTER MODEL (LECTURE 17)

The Friedmann models are usually discussed with respect to the Robertson-Walker line element

$$ds^2 = dt^2 - Q^2(t)\left[\frac{dr^2}{1 - kr^2} + r^2(d\theta^2 + \sin^2\theta\, d\phi^2)\right]; \qquad k = 0, \pm 1. \quad (1)$$

The spaces $k = 0, \pm 1$, are each conformal to Minkowski space. Because our gravitational theory is conformally invariant, we can evidently consider the Friedmann cosmologies in Minkowski space, provided we take care to transform the masses of particles according to $m^* = \Omega^{-1}m$. The masses are taken as constants with respect to the line element (1), $m = m_0$, say, so that (1) gives the conformal frame in which the gravitational equations reduce to the Einstein form. In this chapter we shall consider the three spaces, $k = 0$, ± 1, in turn.

For $k = 0$, we have the Einstein–de Sitter model, with

$$ds^2 = dt^2 - Q^2(t)[dr^2 + r^2(d\theta^2 + \sin^2\theta\, d\phi^2)]. \quad (2)$$

The line element (2) is easily seen to be conformal to Minkowski space by defining a new time-coordinate

$$\tau = \int_0^t \frac{dt}{Q}, \tag{3}$$

giving

$$ds^2 = Q^2(t)[d\tau^2 - dr^2 - r^2(d\theta^2 + \sin^2\theta \, d\phi^2)]. \tag{4}$$

Choose the conformal function to be

$$\Omega = Q^{-1}. \tag{5}$$

The corresponding starred geometry is then

$$ds^{*2} = d\tau^2 - dr^2 - r^2(d\theta^2 + \sin^2\theta \, d\phi^2), \tag{6}$$

and the particle masses are

$$m^* = Qm_0. \tag{7}$$

Write n for the particle density in the original Robertson-Walker frame. The corresponding density n^* in the Minkowski frame is

$$n^* = \Omega^{-3}n = Q^3 n = \text{constant}, \tag{8}$$

the product $Q^3 n$ being constant in all Friedmann cosmologies.

The particles have spatial coordinates r, θ, ϕ, independent of t in the Robertson-Walker frame, and hence independent of τ in the Minkowski frame. Writing L^{-3} for the constant density in the Minkowski frame, we have

$$n^* = Q^3 n = L^{-3}. \tag{9}$$

The Einstein–de Sitter cosmology expressed in Minkowski space is simply a uniform static distribution of particles.

Next we determine Q in terms of τ, so that (7) can be used to give the time-dependence in Minkowski space of the particle masses. The gravitational equations in Robertson-Walker space (i.e., the Einstein equations) give

$$\frac{\dot{Q}^2}{Q^2} = \frac{8\pi G}{3}\sigma, \tag{10}$$

where $\dot{Q} = dQ/dt$ and σ is the mass density. Writing $\sigma = nm_0$ and remembering the definition

$$8\pi G = 6\lambda^2/m_0^2, \tag{11}$$

(10) can be written in the form

$$\dot{Q}^2 = \frac{2\lambda^2}{m_0}\frac{nQ^3}{Q} = \left(\frac{2\lambda^2 L^{-3}}{m_0}\right)\frac{1}{Q}, \tag{12}$$

the first factor on the righthand side of (12) being a constant. Integrating (12) subject to $Q = 0$ at $t = 0$ gives

$$Q(t) = \left(\frac{2\lambda^2 L^{-3}}{m_0}\right)^{1/3}\left(\frac{3}{2}t\right)^{2/3}. \tag{13}$$

Substituting (13) in (3) then leads to the following relation between τ and t;

$$\frac{1}{2}\tau = \left(\frac{2\lambda^2 L^{-3}}{m_0}\right)^{-1/3}\left(\frac{3}{2}t\right)^{1/3}. \tag{14}$$

Replacing t by τ in (13), we have

$$Q(t) = \frac{1}{2}\frac{\lambda^2 L^{-3}}{m_0}\tau^2, \tag{15}$$

whence

$$m^*(\tau) = \frac{1}{2}\lambda^2 L^{-3}\tau^2. \tag{16}$$

The masses in the Minkowski frame are quadratically dependent on the time.

Since we are working with a conformally invariant theory, all relations between theory and observation in the Minkowski frame must be the same as in the Robertson-Walker frame. Consider the Hubble redshift-magnitude relation.

By a standard galaxy we mean one with a specified number of stars, each with the same number of particles of the same kind. Light from a galaxy with coordinate r must start its journey at time $\tau - r$ in order to reach an observer at the spacetime point $r = 0$ at time τ. Write $L^*(\tau - r)$ for the intrinsic luminosity of the galaxy at that time. The observed flux is just the Euclidean value

$$\frac{L^*(\tau - r)}{4\pi r^2}. \tag{17}$$

Now the intrinsic luminosities of standard galaxies are usually taken to have the same value, say, L, in the Robertson-Walker frame, where the particle masses are also the same. In the Minkowski frame, on the other hand, we must have

$$L^*(\tau - r) = [\Omega(\tau - r)]^{-2}L = Q^2(\tau - r)L, \tag{18}$$

because "luminosity" being "energy per unit time" behaves in the same way as (mass)^2 under a conformal transformation. From (15) we therefore see that the observed flux from a galaxy at distance r depends on r according to

$$\frac{(\tau - r)^4}{r^2}. \tag{19}$$

Since the redshift $1 + z$ is given by

$$1 + z = \frac{m^*(\tau)}{m^*(\tau - r)} = \left(\frac{\tau}{\tau - r}\right)^2, \tag{20}$$

it follows, by eliminating r between (19) and (20), that the observed flux depends on z according to

$$\frac{1}{1 + z}\frac{1}{(\sqrt{1 + z} - 1)^2}, \tag{21}$$

which is the usual Hubble relation for the Einstein–de Sitter model.

The dependence (16) of mass on time was obtained above by starting with $m = m_0 =$ constant in the Robertson-Walker frame. Can we also obtain (16) by direct calculation? In the Minkowski frame, the time-symmetric mass propagator takes the simple form

$$\widetilde{G}(X, B) = \frac{1}{4\pi}\,\delta(s_{XB}{}^{*2}), \tag{22}$$

while the fully retarded mass propagator is

$$G^{\text{ret}}(X, B) = \frac{1}{4\pi}\,\delta(t_X - t_B - |\boldsymbol{x}_X - \boldsymbol{x}_B|). \tag{23}$$

Considering the retarded mass field generated by (23), and accepting that the universe came into being at $t = 0$ in the Robertson-Walker frame, and hence at $\tau = 0$ in the Minkowski frame, only particles at distances $<\tau_X$. contribute to $m^{*(\text{ret})}(X)$. The contribution of such a particle at distance r is simply

$$\frac{\lambda^2}{4\pi r}, \tag{24}$$

and the total contribution for a uniform distribution of particles of density L^{-3} is

$$\frac{\lambda^2 L^{-3}}{4\pi} \int_0^\pi \sin\theta \, d\theta \int_0^{2\pi} d\phi \int_0^{\tau_X} r \, dr = \frac{1}{2} \lambda^3 L^{-3} \tau_X^2, \tag{25}$$

which is the same as (16). Hence (16) can be obtained by direct calculation.

The Space for $k = +1$

Defining $\sin R = r$, we have

$$\begin{aligned} ds^2 &= dt^2 - Q^2(t)\left[\frac{dr^2}{1 - r^2} + r^2(d\theta^2 + \sin^2\theta \, d\phi^2)\right] \\ &= dt^2 - Q^2(t)[dR^2 + \sin^2 R(d\theta^2 + \sin^2\theta \, d\phi^2)]. \end{aligned} \tag{26}$$

The coordinate r is to be regarded as dimensionless, whereas Q has dimensionality C (see §5.1). From the Einstein equation we get

$$\frac{\dot{Q}^2}{Q^2} + \frac{1}{Q^2} = \frac{8\pi G}{3}\sigma = \frac{8\pi G}{3} n m_0, \tag{27}$$

where $n(t)$ is the particle density. Noting that the dimensionless quantity nQ^3 is independent of t, we can write

$$\dot{Q}^2 = \frac{\alpha - Q}{Q}, \tag{28}$$

where the constant α with dimensionality C is given by

$$\alpha = \frac{8\pi G}{3} m_0 \cdot nQ^3 = \frac{2\lambda^2}{m_0} nQ^3. \tag{29}$$

Equation (28) can be solved in terms of a dimensionless quantity T, defined by

$$T = \int_0^t \frac{dt}{Q}, \tag{30}$$

to give

$$Q = \frac{\alpha}{2}(1 - \cos T), \tag{31}$$

$$t = \frac{\alpha}{2}(T - \sin T). \tag{32}$$

In terms of T, (26) becomes

$$ds^2 = Q^2[dT^2 - dR^2 - \sin^2 R(d\theta^2 + \sin^2\theta\, d\phi^2)]. \tag{33}$$

Next it can be verified that the *coordinate* transformations

$$\xi = \frac{1}{2}(T + R), \qquad X = \tan\xi, \qquad 2\tau = X + Y,$$

$$\eta = \frac{1}{2}(T - R), \qquad Y = \tan\eta, \qquad 2\rho = X - Y, \tag{34}$$

map the interior of the triangle formed in the R, T, plane by the points $(0, 0)$, $(\pi, 0)$, $(0, \pi)$, into the quadrant $0 \leq \rho < \infty$, $0 \leq \tau < \infty$, in the ρ, τ plane. By carrying through these transformations, it can also be shown that

$$ds^2 = \frac{4Q^2}{(1 + X^2)(1 + Y^2)}[d\tau^2 - d\rho^2 - \rho^2(d\theta^2 + \sin^2\theta\, d\phi^2)]. \tag{35}$$

Therefore the *conformal* transformation

$$\Omega = C(1 + X^2)^{1/2}(1 + Y^2)^{1/2}/2Q \tag{36}$$

maps the original spherical space into half of the Minkowski space

$$ds^{*2} = C^2[d\tau^2 - d\rho^2 - \rho^2(d\theta^2 + \sin^2\theta\, d\phi^2)]; \qquad 0 \leq \tau < \infty. \tag{37}$$

To keep track of dimensionalities, the basic unit C has been introduced into (36), so that Ω is dimensionless. Numerically, C is unity. The factor C^2 appears on the righthand side of (37) because τ, ρ, are dimensionless.

Since the total mass $m = m_0$ was constant in the Robertson-Walker frame, the mass m^* is variable in the Minkowski frame and is given by

$$m^* = \frac{2Qm\, C^{-1}}{(1 + X^2)^{1/2}(1 + Y^2)^{1/2}}$$

$$= \frac{[1 + (\tau + \rho)^2]^{1/2}[1 + (\tau - \rho)^2]^{1/2} + \tau^2 - \rho^2 - 1}{[1 + (\tau + \rho)^2][1 + (\tau - \rho)^2]} \alpha\, C^{-1}m. \quad (38)$$

The particle trajectories are $R =$ constant in the Robertson-Walker frame and are given by

$$\frac{2\rho}{1 + \tau^2 - \rho^2} = \text{constant} \qquad (A^{-1}, \text{ say}), \quad (39)$$

in the Minkowski frame. In the Robertson-Walker frame, the particle density is independent of R. In the Minkowski frame, on the other hand, the density depends on both ρ and τ. At $\tau = 0$, it has the form

$$(1 + \rho^2)^{-3}. \quad (40)$$

Thereafter it changes according to the trajectories (39). In terms of the constant A, the radial coordinate along a trajectory is

$$\rho = (A^2 + \tau^2 + 1)^{1/2} - A, \quad (41)$$

from which we see that $d\rho/d\tau = 0$ at $\tau = 0$, and that $\rho \to \tau$ as $\tau \to \infty$.

How shall we interpret these results? At $\tau = 0$ we have a spherically symmetric cloud with the radial density distribution (40). Initially, at $\tau = 0$, the cloud is at rest, but as τ increases, the cloud moves radially outwards and ultimately dissipates itself. It does so, not because of gravitational forces, which are zero in this frame, but because $\partial m^*/\partial\tau$, $\partial m^*/\partial\rho$, are nonzero, and these mass-gradient components appear in the equations of motion of the particles, i.e., in equations (138) of Chapter 6.

It would be possible to derive (38) by direct calculation. Likewise it would be possible to derive the usual Hubble redshift-magnitude relation for the case $k = +1$, just as we did for the Einstein–de Sitter model.

It follows from (38) that m^* is zero at $\tau = 0$, increases to a maximum as τ increases, and then again tends to zero as $\tau \to \infty$. This result means that m^* is initially zero because mass interactions are restricted to particles at distances $r < \tau$, which gives $m^* = 0$ at $\tau = 0$, and that $m^* \to 0$ as $\tau \to \infty$ because the finite cloud dissipates itself and the mass interactions tend to zero.

It comes as no little surprise to find the cloud dispersing to infinity. In the Robertson-Walker frame, we think of the universe in the $k = +1$ case as expanding from a singularity to a maximum of the function $Q(t)$ and then collapsing back into a singularity. We think of the collapse as being a reverse of the expansion, expansion and contraction being determined by the two roots of (27). How does this reverse express itself in Minkowski space?

From (31) we see that Q attains its maximum value for $T = \pi$. Yet the dispersal of the cloud to infinity in the Minkowski frame corresponds to the behavior of the system in the triangle $(0, 0)$, $(\pi, 0)$, $(0, \pi)$, in the T, R, plane; so the dispersal corresponds only to the expansion phase in the Robertson-Walker frame. To obtain the contraction phase in which the dispersed cloud at infinity falls back to its original state at $\tau = 0$, one requires the range π to 2π for the T coordinate. The triangles in the T, R, plane with this range, and which map into the upper half of the Minkowski plane, are

$$(\pi, \pi + 2n\pi), \qquad (2\pi, \pi + 2n\pi), \qquad (\pi, 2\pi + 2n\pi), \tag{42}$$

with n any integer. The periodicity of R with respect to π is not unexpected, but the addition of π to R, compared to the expansion phase, in which the possible triangles in the T, R, plane with $0 \leq T \leq \pi$ are

$$(0, 2n\pi), \qquad (\pi, 2n\pi), \qquad (0, \pi + 2n\pi), \tag{43}$$

requires a change of sign in the relation between R and the original r coordinate, to $r = -\sin R$. The metric (26) is clearly unchanged when π is added to R.

The corresponding triangles going into the lower half of the Minkowski plane are

$$(0, \pi + 2n\pi), \qquad (\pi, \pi + 2n\pi), \qquad (\pi, 2\pi + 2n\pi), \tag{44}$$

for the range 0 to π for T, and

$$(\pi, 2n\pi), \qquad (2\pi, 2n\pi), \qquad (2\pi, \pi + 2n\pi), \tag{45}$$

for $\pi \leq T \leq 2\pi$. The situation for the triangles (42), (43), (44), (45), is shown in Figure 7.1. In terms of the original t-coordinate, the time required for the dispersal of the cloud—i.e., the expansion phase in the Robertson-Walker frame—is given by putting $T = \pi$ in (32), and is $\frac{1}{2}\pi\alpha$.

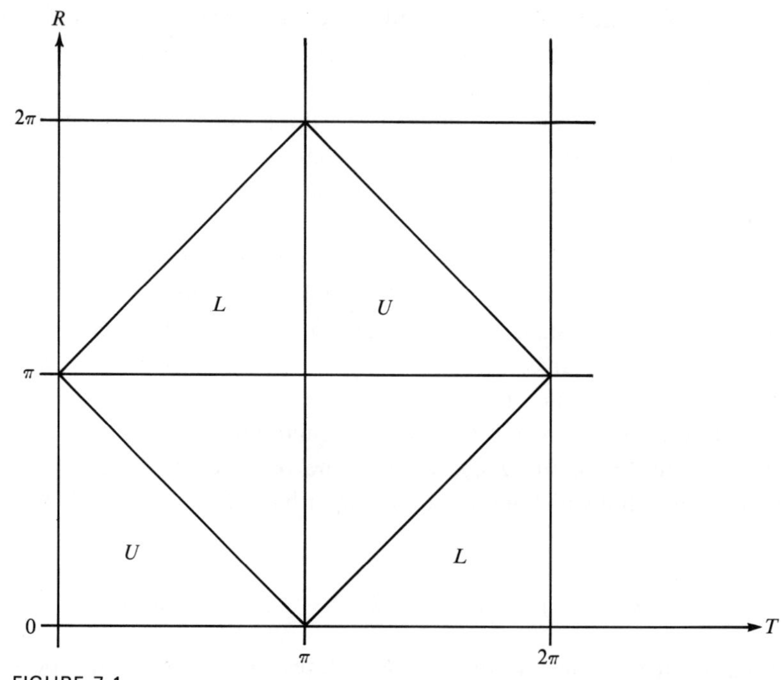

FIGURE 7.1
The triangles marked U are mapped into the upper half of the Minkowski plane, those marked L into the lower half. The blank triangles have no mapping into the Minkowski plane, since they correspond to $\rho \leq 0$. The diagram repeats itself with periodicity 2π with respect to both T and R.

The Space for $k = -1$

Define $\sinh R = r$, so that the metric (1) for this space can be written in the form

$$ds^2 = dt^2 - Q^2(t)[dR^2 + \sinh^2 R(d\theta^2 + \sin^2\theta\, d\phi^2)]. \tag{46}$$

From the gravitational equations we have

$$\dot{Q}^2 = \frac{\alpha + Q}{Q}, \tag{47}$$

with α defined as in (29). Once again (47) can be solved in terms of a dimensionless T defined by (30), to give

$$Q = \frac{\alpha}{2}(\cosh T - 1), \tag{48}$$

$$t = \frac{\alpha}{2}(\sinh T - T). \tag{49}$$

The appropriate coordinate transformations differ from (34) in the definitions of X, Y,

$$X = \tanh \xi, \qquad Y = \tanh \eta, \tag{50}$$

but are otherwise the same. The coordinate transformations now map the quadrant $0 \leq R < \infty$, $0 \leq T < \infty$, in the R, T, plane into the interior of the triangle $(0,0)$, $(1,0)$, $(0,1)$, in the ρ, τ, plane. Corresponding to (35) we have

$$ds^2 = \frac{4Q^2}{(1 - X^2)(1 - Y^2)}[d\tau^2 - d\rho^2 - \rho^2(d\theta^2 + \sin^2\theta\, d\phi^2)]. \tag{51}$$

The function required for conformal transformation to Minkowski space is therefore

$$\Omega = [C(1 - X^2)^{1/2}(1 - Y^2)^{1/2}]/2Q, \tag{52}$$

and the resulting starred geometry is

$$ds^{*2} = C^2[d\tau^2 - d\rho^2 - \rho^2(d\theta^2 + \sin^2\theta\, d\phi^2)]. \tag{53}$$

Corresponding to (38), (39), we have

$$m^* = \frac{2Q\, C^{-1}m}{(1 - X^2)^{1/2}(1 - Y^2)^{1/2}} \tag{54}$$

$$= \frac{1 + \tau^2 - \rho^2 - [1 - (\tau + \rho)^2]^{1/2}[1 - (\tau - \rho)^2]^{1/2}}{[1 - (\tau + \rho)^2][1 - (\tau - \rho)^2]}\alpha C^{-1}m, \tag{55}$$

$$\frac{2\rho}{1 + \rho^2 - \tau^2} = \text{constant} \qquad (A^{-1}, \text{ say}), \tag{56}$$

and corresponding to (40), (41), we have

$$(1 - \rho^2)^{-3}, \tag{57}$$

$$\rho = A - (A^2 + \tau^2 - 1)^{1/2}. \tag{58}$$

We again have $d\rho/d\tau = 0$ at $\tau = 0$, but the radial coordinate ρ now tends to zero as $\tau \rightarrow 1$.

The $k = -1$ Friedmann model is therefore to be interpreted as an initially stationary cloud confined to $\rho \leq 1$. It collapses to zero radius at $\tau = 1$, and it does so because of an initial singularity in the density of matter (57) as the radial coordinate approaches 1. Once again the collapse occurs because of mass-gradient components in the equations of motion (138) of Chapter 6.

§7.2 THE ORIGIN OF THE UNIVERSE

Throughout the preceding discussion, we took as our starting point the fact that the various Friedmann models are known to satisfy the gravitational equations. The latter are usually considered in terms of a smooth-fluid approximation in which a density function $\rho(X)$ is defined by

$$\rho(X) = \int_{\substack{\text{unit [3]} \\ \text{proper volume}}} \sum_a \frac{\delta_4(X, A)}{\sqrt{-g(A)}} m(A)\, da, \tag{59}$$

it being assumed that the length unit is large enough for many terms in Σ_a to contribute to ρ. In a general conformal frame, the gravitational equations are given by (117) of Chapter 6. For the line element (1), and for the Einstein–de Sitter model, these equations lead to two independent conditions

$$\frac{\dot{Q}^2}{Q^2} = \frac{2\rho}{m^2} - \frac{\dot{m}^2}{m^2} - \frac{2\dot{Q}}{Q}\frac{\dot{m}}{m}, \tag{60}$$

$$\frac{2\ddot{Q}}{Q} + \frac{\dot{Q}^2}{Q^2} = -\frac{4\dot{m}}{m}\frac{\dot{Q}}{Q} - \frac{2\ddot{m}}{m} + \frac{\dot{m}^2}{m^2}, \tag{61}$$

where the dots denote differentiation with respect to time, and where for convenience we are putting the coupling constant $\lambda = 1$. Because of the spatial homogeneity of the model, only the time derivatives of the mass function m appear in the smooth-fluid approximation.

In the conformal frame in which m is constant, (61) leads to $Q(t) \propto t^{2/3}$, whereas in the conformal frame in which $Q = 1$ at all times, the same equation gives $m \propto \tau^2$. We have seen above how this latter proportionality can be obtained in a smooth-fluid calculation. We end the present chapter by considering $m \propto \tau^2$ in relation to the discrete-particle picture in which

$$m(X) = \sum_a \int G^{(\text{ret})}(X, A)\, da. \tag{62}$$

The particles are static in the Minkowski conformal frame and

$$G^{(\mathrm{ret})}(X, A) = \frac{1}{4\pi|\boldsymbol{x}_X - \boldsymbol{x}_A|}\,\delta(\tau_X - \tau_A - |\boldsymbol{x}_X - \boldsymbol{x}_A|). \tag{63}$$

Thus we can write

$$m(X) = \frac{1}{4\pi}\sum_a \frac{1}{|\boldsymbol{x}_X - \boldsymbol{x}_A|}\int \delta(\tau_X - \tau_A - |\boldsymbol{x}_X - \boldsymbol{x}_A|)\,d\tau_A, \tag{64}$$

and

$$\begin{aligned} \dot{m} &= \frac{1}{4\pi}\sum_a \frac{1}{|\boldsymbol{x}_X - \boldsymbol{x}_A|}\int \left[\frac{d}{d\tau_X}\,\delta(\tau_X - \tau_A - |\boldsymbol{x}_X - \boldsymbol{x}_A|)\right] d\tau_A \\ &= -\frac{1}{4\pi}\sum_a \frac{1}{|\boldsymbol{x}_X - \boldsymbol{x}_A|}\int \left[\frac{d}{d\tau_A}\,\delta(\tau_X - \tau_A - |\boldsymbol{x}_X - \boldsymbol{x}_A|)\right] d\tau_A. \end{aligned} \tag{65}$$

The particle paths are considered to go from $\tau = 0$ to $\tau = \infty$, so that

$$\dot{m} = \frac{1}{4\pi}\sum_a \frac{1}{|\boldsymbol{x}_X - \boldsymbol{x}_A|}\,\delta(\tau_X - |\boldsymbol{x}_X - \boldsymbol{x}_A|), \tag{66}$$

the delta function being zero for $\tau_A \to \infty$. If the paths were considered to go from $-\infty$ to ∞, the delta functions would be zero at both "ends," and $\dot{m}$ would be zero. Although (61) would still be satisfied, (60) would then lead to $\rho = 0$. Hence we see that it is the broken ends of the paths at $\tau = 0$ which permit a nontrivial model to exist. Writing $r = |\boldsymbol{x}_X - \boldsymbol{x}_A|$, dropping the subscript X, and evaluating (66) in integral form, we have

$$\dot{m}(\tau) = n\int_0^\infty \delta(\tau - r)r\,dr = n\tau, \tag{67}$$

where n is the proper density of the particles.[1] This is the same result for $\dot{m}$ as before.

The fact that broken ends thus play a critical role in the Einstein–de Sitter model, and similarly for the models with $k = \pm 1$, means that the usual forms of the Friedmann cosmologies are not in principle different from the

[1] Owing to the sense of the discussion here, we have dropped the starred notation for the Minkowski frame. The n used here was written earlier in the chapter as n^* and m was written as m^*.

steady-state model in which broken paths represent particle creation. It is true that broken ends appear only at $\tau = 0$ for the Friedmann models, whereas broken ends appear at all times in the steady-state model, but this difference is technical rather than one of principle.

However, in Chapter 9 we shall see how to avoid broken paths entirely, and we shall find that the concept of an "origin" for the universe can be replaced by a new form of symmetry.

8

PROBLEMS IN THE QUANTIZATION OF INERTIA AND GRAVITATION

§8.1. THE FREE-PARTICLE PROPAGATOR (LECTURE 18)

In §4.6 we saw how to build the propagator for free Dirac particles from the infinitesimal propagator

$$K(X+Q;X) = \frac{1}{\pi}\left[\not{q}\,\delta'(q^2) - \frac{1}{2}\,\mathrm{im}\,\delta(q^2)\right], \tag{1}$$

where q^i is the coordinate displacement between point $X+Q$ and point X, $q^2 = q_i q^i$, and $\not{q} = \gamma_i q^i$, γ_i being the Dirac matrices. For a path Γ^+ going forward in time from point 1 to point 2, define an amplitude by

$$P(\Gamma^+) = \lim_{N\to\infty} \prod_{r=1}^{N} A_r\, K(Q_r, Q_{r-1}), \tag{2}$$

as in (88) of Chapter 40. Here we have identified the point Q_0 with point 1 and Q_N with point 2; $Q_1, \ldots, Q_{N-1}$ are intermediate points chosen so that the steps Q_{r-1} to Q_r are small. With a suitable choice for the measure factors A_r, the path integral

$$\int P(\Gamma^+)\mathscr{D}\Gamma^+ \tag{3}$$

gives

$$\begin{aligned} K_0^+(2;1) &= \theta(t_2 - t_1) \sum_n u_n(2)\,\bar{u}_n(1) \\ &= \frac{1}{2\pi}\theta(t_2 - t_1)(\nabla\!\!\!/_2 - im)\left[\delta(s_{21}{}^2) - \frac{m}{2s_{21}}\theta(s_{21}{}^2)J_1(ms_{21})\right], \end{aligned} \tag{4}$$

where u_n are the free-particle solutions of the Dirac equation. The same infinitesimal propagator applied to paths Γ^- going backward in time gives

$$\begin{aligned} K_0^-(2;1) &= -\theta(t_1 - t_2) \sum_n u_n(2)\,\bar{u}_n(1) \\ &= \frac{1}{2\pi}\theta(t_1 - t_2)(-\nabla\!\!\!/_2 + im)\left[\delta(s_{21}{}^2) - \frac{m}{2s_{21}}\theta(s_{21}{}^2)J_1(ms_{21})\right]. \end{aligned} \tag{5}$$

The situation for the path integrals leading to (4), (5), is different from the nonrelativistic case in several respects. The infinitesimal propagator is now a 4×4 matrix. The product in (2) is a chain of 4×4 matrices in which adjacent members are in general noncommutative. In the nonrelativistic case, on the other hand, we have $\exp(idS)$ for the amplitude associated with a path element of action dS. The elementary amplitudes for a chain of such steps multiply as numbers to give $\exp(iS)$ for the finite path amplitude, leading to

$$\int \exp[iS(\Gamma_{21})]\mathscr{D}\Gamma_{21} \tag{6}$$

for the nonrelativistic propagator. In this case the amplitude for a finite path has the same form as that for an infinitesimal step, namely, an exponential with the action as its phase. This property is lost in the relativistic case, in the sense that the finite propagator, whether K_0^+, K_0^-, or $\frac{1}{2}(K_0^+ + K_0^-)$, is different from the infinitesimal propagator, not only in appearance, but in the important physical respect that, whereas the infinitesimal propagator from point 1 is zero except on the light cone of point 1, the finite propagator is nonzero off this light cone. The Bessel function J_1 in (4), (5), is nonzero inside the future and past light cones of point 1, respectively. It is interesting to consider this difference in terms of the Dirac equation for the finite propagator,

$$(\nabla\!\!\!/_2 + im)K(2;1) = \delta_4(2;1), \tag{7}$$

rather than in terms of path integrals.

All forms of the finite propagator, K_0^+, K_0^-, $\frac{1}{2}(K_0^+ + K_0^-)$, K_+, or K_-, (a propagator like K_+, but with the reversed time sense) satisfy (7). Hence (7) alone does not specify the propagator uniquely. The situation here is analogous to the fact that the potential in electromagnetic theory is not determined uniquely by the wave equation.

Let $K^{(0)}(2;1)$ be chosen to satisfy

$$\nabla\!\!\!/_2 K^{(0)}(2;1) = \delta_4(2,1), \tag{8}$$

and let $K^{(n)}(2;1)$ be chosen so that

$$\nabla\!\!\!/_2 K^{(n)}(2;1) = -im\, K^{(n-1)}(2;1). \tag{9}$$

Then

$$K(2;1) = \sum_{n=0}^{\infty} K^{(n)}(2;1) \tag{10}$$

satisfies (7). Since

$$I(2;1) = \frac{1}{4\pi}\,\delta(s_{21}{}^2) \tag{11}$$

has the property

$$\nabla\!\!\!/_2 \cdot \nabla\!\!\!/_2\, I(2;1) = \Box I(2;1) = \delta_4(2,1), \tag{12}$$

it is clear that the choice

$$K^{(0)}(2;1) = \nabla\!\!\!/_2\, I(2;1) \tag{13}$$

meets the requirement (8). Comparing (8) and (9), we see that $K^{(0)}$ is a Green's function for the determination of $K^{(n)}$ in terms of $K^{(n-1)}$,

$$K^{(n)}(2;1) = -im \int K^{(0)}(2;3)K^{(n-1)}(3;1)\, d\tau_3, \tag{14}$$

the integral being over all spacetime. Reducing $K^{(n-1)}$ in the same way, and so on, we get

$$K^{(n)}(2;1) = (-im)^n \int K^{(0)}(2;2+n)K^{(0)}(2+n;1+n)\ldots$$
$$K^{(0)}(3;1)\, d\tau_3 \ldots d\tau_{2+n}. \tag{15}$$

The choice (11) for $I(2; 1)$, being time-symmetric, leads to the symmetric form

$$\frac{1}{2}[K_0^+(2;1) + K_0^-(2;1)] \tag{16}$$

for $K(2; 1)$. If in (11) we had used $\delta_+(s_{21}{}^2)$ instead of $\delta(s_{21}{}^2)$, the same equations (12) to (15) would have given

$$K(2;1) = K_+(2;1), \tag{17}$$

and $\delta_-(s_{21}{}^2)$ in (11) would have given K_-.

For $n = 1$ we could already have chosen

$$K^{(1)}(2;1) = -im\, I(2;1), \tag{18}$$

since (9) for $n = 1$ is then satisfied identically. The form (18) can also be obtained from (14). Thus

$$\begin{aligned} K^{(1)}(2;1) &= -im \int K^{(0)}(2;3)K^{(0)}(3;1)\, d\tau_3 = -im \int \nabla\!\!\!\!/_2 I(2;3)\, \nabla\!\!\!\!/_3 I(3;1)\, d\tau_3 \\ &= im \int \nabla\!\!\!\!/_3 I(2;3)\, \nabla\!\!\!\!/_3 I(3;1)\, d\tau_3 = -im \int I(2;3)\, \Box I(3;1)\, d\tau_3 \\ &= -im\, I(2;1), \end{aligned} \tag{19}$$

a surface integral at infinity being set equal to zero.

Except for a factor $\frac{1}{2}$ [which leads to the factor $\frac{1}{2}$ in (16)], the first two terms of (10) are the same as the two terms in the infinitesimal propagator (1). The first two terms of (10) contribute only when the displacement from point 1 to point 2 is null, but higher-order terms contribute for nonnull displacements. It is these higher-order terms in (10) that generate the Bessel functions in (4), (5). Indeed, if the Bessel functions are expanded in powers of m, the terms are twice the corresponding powers of m in (10).

We can think of (15) as a summation of all paths with n corners. The particle goes from point 1 to point 3 at $d\tau_3$, then to point 4 at $d\tau_4, \ldots,$ to point $2 + n$ at $d\tau_{2+n}$, and finally to point 2. The segments of such a path are all null. Since $K^{(0)}$ is the propagator for a massless spin-$\frac{1}{2}$ particle (we can call such a particle a neutrino), we see that a particle with mass m going from 1 to 2 typically goes from 1 to some point 3 like a neutrino. At 3, unlike a neutrino, it is scattered with amplitude $-im$, whereon it proceeds like a neutrino to point 4, and so on until it arrives at 2. At all the points $3, 4, \ldots, 2 + n$, it experiences scattering, as is shown in Figure 8.1. Evidently

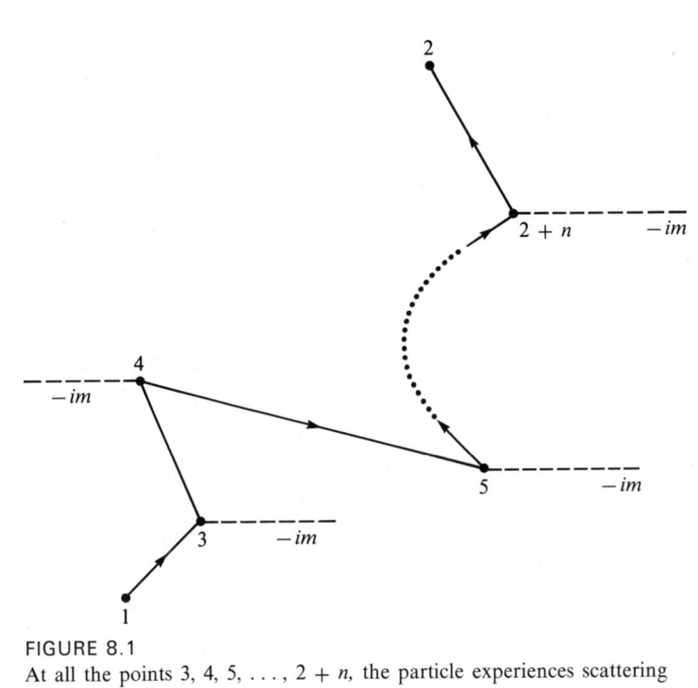

FIGURE 8.1
At all the points 3, 4, 5, . . . , $2 + n$, the particle experiences scattering with amplitude $-im$.

we have a situation analogous to the perturbation expansion of electrodynamics, the terms in (10) giving the different orders of the expansion. In place of the electric charge, we now have the mass appearing in the scattering amplitude.

This description of the free-particle propagator suggests that the phenomenon of "mass" has the character of a scattering field, a concept which agrees with the classical picture developed in the preceding chapters. To be consistent with the point of view developed for electrodynamics, we require the symmetric form (16) to represent the local propagator, not the time-asymmetric forms K_0^+, K_0^-, K_+, or K_-. The above work shows that (16) can be represented by the expansion (10) with $K^{(0)}$ given by (11), (13), and with $K^{(n)}$, $n \geq 1$, given by (15). Since the integrations $d\tau_3$, $d\tau_4$, . . . , in (15) are over all spacetime, there is no restriction concerning time-reversals on the paths which appear in (15). However, the integrations give rise to extensive cancelations among the paths. The forms (4), (5), for K_0^+, K_0^-, show that (16) is zero when point 2 is outside the light cone of point 1, a property which develops when the integrations are carried through explicitly. Indeed, the paths of free particles with time-reversals are self-canceling, a property with which we are familiar from Chapter 4. This same property is shown by the fact that K_0^+ can be obtained from a path integral with respect to Γ^+ paths, and K_0^- can be obtained with Γ^- paths—no time-reversals occurring in either case.

We come now to a critical issue. The integrations over $d\tau_3$, $d\tau_4$, . . . , in (15) cover all spacetime. For this situation to be correct, we have to be sure the "mass field" m is everywhere the same throughout the universe, and also that the particle in question is not subject to a process in which it is changed into some other kind of particle. Since we cannot be sure of these things, we proceed by restricting the integrations over $d\tau_3$, $d\tau_4$, . . . , to a finite region, say, V, of spacetime which is nevertheless large compared to the local system in question. The integrations cover all paths that go from 1 to 2 without leaving V. However, there will now be paths from 1 that go outside V. Those paths that reach 2 as a consequence of paths going outside V from 1 lead to what we might call a "response of the universe," in analogy to the situation in electrodynamics. The response can be written quite generally as the sum of a part which extends the integrations $d\tau_3$, $d\tau_4$, . . . , in (15) to cover the whole of spacetime, so that $\frac{1}{2}\cdot(K_0^+ + K_0^-)$ again forms a part of the total propagator, and an as-yet unknown response propagator, say, $\kappa(2; 1)$, satisfying the homogeneous Dirac equation

$$(\nabla\!\!\!\!/_2 + im)\,\kappa(2; 1) = 0 \tag{20}$$

inside V. We have

$$K(2; 1) = \frac{1}{2}[K_0^+(2; 1) + K_0^-(2; 1)] + \kappa(2; 1). \tag{21}$$

We carry the analogy with electrodynamics still further by adopting the notation:

$$\begin{aligned} K(2; 1)_{\text{ret}} \equiv K_0^+(2; 1) &= \theta(t_2 - t_1)\Big[\sum_{E_n>0} u(2)\bar{u}(1) + \sum_{E_n<0} u(2)\bar{u}(1)\Big] \\ &= K(2; 1)_{\text{ret}+} + K(2; 1)_{\text{ret}-}, \end{aligned} \tag{22}$$

$$\begin{aligned} K(2; 1)_{\text{adv}} \equiv K_0^-(2; 1) &= -\theta(t_1 - t_2)\Big[\sum_{E_n>0} u(2)\bar{u}(1) + \sum_{E_n<0} u(2)\bar{u}(1)\Big] \\ &= K(2; 1)_{\text{adv}+} + K(2; 1)_{\text{adv}-}. \end{aligned} \tag{23}$$

Here we have simply separated K_0^+, K_0^-, into positive- and negative-frequency parts. We now take

$$\kappa(2;1) = \frac{1}{2}[(K_{\text{ret}+} - K_{\text{adv}+}) + (K_{\text{adv}-} - K_{\text{ret}-})]. \tag{24}$$

Again notationally, if we define a time-inverted propagator K' by

$$K'_{\text{ret}} = K_{\text{adv}}, \qquad K'_{\text{adv}} = K_{\text{ret}}, \tag{25}$$

so that (24) can be written as

$$\kappa(2;1) = \frac{1}{2}[K_{\text{ret}+} - K_{\text{adv}+}) + (K'_{\text{ret}-} - K'_{\text{adv}-})], \tag{26}$$

we have exactly the same form as the electrodynamic response condition given in (124) of Chapter 4. [In the latter condition, the potentials A_i come from paths and A'_i come from conjugate paths. In the electromagnetic action, the line integrals involving A_i and A'_i have opposite signs, which implies opposite time-sense.]

Substituting in (24) from the identifications in (22), (23), we get immediately

$$\kappa(2;1) = \frac{1}{2}\left[\sum_{E_n>0} u_n(2)\bar{u}_n(1) - \sum_{E_n<0} u_n(2)\bar{u}_n(1)\right], \tag{27}$$

irrespective of the time-order of points 1 and 2. Inserting (27) in (21) then gives the important result

$$K(2;1) = K_+(2;1). \tag{28}$$

Our point of view is that, although $\frac{1}{2}(K_0^+ + K_0^-)$ is the local propagator for free Dirac particles, the response of the universe added to the local propagator gives K_+. *This result now replaces the hypothesis of Chapter 4 in which we assumed the positive-energy states of a free particle to be propagated by K_0^+ and the negative-energy states by K_0^-.*

The propagator $K_+(2;1)$ falls off as $\exp(-ms_{21})$ outside the light cone of point 1. This extension of K_+ outside the light cone is usually regarded as a small effect. But with path integrals, we are concerned with steps much smaller than m^{-1}. Indeed, we have to think of $s_{21} \simeq m^{-1}$ as a very large distance. The K_+ propagator thus permits a particle to arrive at point 2 even though the displacement from 1 to 2 is grossly spacelike. Unless we abandon our physical concepts in a very major respect, the particle cannot have come directly from point 1. It must have arrived via the universe.

As an aside, it is worth noting that the perturbation expansion (97) of

Chapter 4 is obtained in Appendix 3 by a method which uses ${K_0}^+$ as the local propagator for the forward time-sense, and ${K_0}^-$ for the backward time-sense. It is not hard to show that $\frac{1}{2}({K_0}^+ + {K_0}^-)$ leads to the same expansion. However, one must not use K_+ in calculating such perturbations, since only local paths are involved in a local electrodynamic perturbation—the use of K_+ would involve electromagnetic integrals over paths going far away into the distant universe. The propagator K_+ must only be used in such a calculation when the particle is free.

§8.2. MASS INTERACTIONS IN QUANTUM MECHANICS (LECTURE 19)

To the first order in the mass, we can write the infinitesimal propagator (1) in the form

$$K(X + Q; X) = \frac{1}{\pi} \not{q}\, \delta'(q^2) \exp\left(\frac{1}{2}\, im\not{q}\right)$$
$$= \frac{1}{2\pi} \exp\left(\frac{1}{2}\, im\not{q}\right) \not{\nabla}_Q\, \delta(q^2), \tag{29}$$

since to this order

$$\exp\left(\frac{1}{2}\, im\not{q}\right) = 1 + \frac{1}{2}\, im\not{q}, \tag{30}$$

and $q^2\, \delta'(q^2) = -\delta(q^2)$. The appearance of the exponential factor in (29) is interesting in relation to the Schrödinger theory, where the amplitude for a path $\Gamma^+{}_{A\tilde{A}}$ of particle a from $\tilde{A}$ to A is

$$\exp\left(-im \int_{\tilde{A}}^{A} da\right) \simeq \exp\left\{-im \int_0^T \left[1 - \frac{1}{2}\, \dot{a}^2(t)\right] dt\right\} \tag{31}$$

to sufficient accuracy, with $t = 0$ at point $\tilde{A}$ and $t = T > 0$ at A. This amplitude cannot be used for Dirac particles because it does not contain spin. It is easy, however, to generalize the lefthand side of (31) to a 4×4 matrix. Working in Minkowski space, from

$$da^2 = \eta_{ik}\, da^i\, da^k,$$
$$\gamma_i \gamma_k + \gamma_k \gamma_i = 2I\eta_{ik}, \tag{32}$$

where γ_i are the Dirac matrices and I is the unit 4×4 matrix, it follows that

$$I \cdot da^2 = (\gamma_i \, da^i)^2. \tag{33}$$

This relation suggests the generalization to

$$\exp\left(im \int_{\tilde{A}}^{A} \gamma_i \, da^i\right)$$

which is similar infinitesimally to the exponential in (29). It is an interesting detail that (29) contains a factor $\frac{1}{2}$ which does not appear in such an attempted generalization of the Schrödinger theory. The major difference lies, however, in the $\not{q}\,\delta(q^2)$ factor in (29), which does not appear at all in a spin generalization of the Schrödinger amplitude. It is just this additional factor that produces the essential difference between nonrelativistic quantum mechanics and the relativistic theory. It is this factor which forces the segments of paths to be null, and which causes most of the difficulty in extending the path-integral method to the relativistic theory.

Must we simply accept the $\not{q}\,\delta(q^2)$ factor in the infinitesimal propagator, or can we understand how it arises in physical terms? Referring back to the Schrödinger amplitude on the lefthand side of (31), we may note that if we let $m \to \infty$, all paths within the light cone of point 1 have infinite action. Because of phase cancelation, such paths make no contribution to the path integral determining the propagator. There would be no timelike classical path determined by $\delta S = 0$ in such a case. The only way to obtain a nonzero propagator would be for all elements of paths contributing to the propagator to be null, just as is in fact required by the $\not{q}\,\delta(q^2)$ factor. Next, we would need to understand the physical meaning of $m \to \infty$. However, we shall refer back to this question in Chapter 9, and turn for the moment to answering a more accessible question.

In §8.1 we found the property of "mass" to show itself in quantum mechanics as a scattering effect analogous to that of the electromagnetic field. The analogy is shown again by the fact that the mass appears in (29) as a multiplying factor in the phase of an exponential—the electromagnetic interaction also appears as an exponential, as we saw in earlier chapters. These are indications that "mass" is to be interpreted as an interaction of a particle with other particles. This was the point of view we took in the classical theory. It is interesting to find the same situation emerging now in quantum mechanics.

The mass interaction on an element da^i of a path of particle a is thus

$\frac{1}{2}m\, d\not{a}$, where $d\not{a} = \gamma_i\, da^i$. This is the total interaction of particle a with all particles in the universe. The elementary interaction with particle b would be

$$\frac{1}{4}\, d\not{a}\, \widetilde{G}(A, B)\, d\not{b}, \tag{34}$$

using the same propagator $\widetilde{G}(A, B)$ as in the classical theory. (Here we may recall that in Chapter 4 the electromagnetic interaction was found to be the same for quantum electrodynamics as for classical electrodynamics.) The interaction (34) relates the element $d\not{a}$ of a path of particle a to an element $d\not{b}$ of a path of particle b, the points A, B, being at $d\not{a}$, $d\not{b}$, respectively. The symmetrization of the interaction with respect to the particle pair a, b, has thus led to a 4×4 matrix appearing both at A and at B. It is important to notice that the γ matrices in $d\not{a} = \gamma_i\, da^i$, $d\not{b} = \gamma_i\, db^i$, are not multiplied. The form (34) is to be interpreted in the sense of an outer product.

At first sight it might seem as if the mass of particle a resulting from summing (34) with respect to all elements of all paths of all particles b would then be a 4×4 matrix. However, when the transition element of (34) is evaluated with respect to the wave function ψ_b of particle b, it is found that the outcome is indeed a scalar contribution to the mass of a. The relevant path integral is worked out in Appendix 5 (p. 256), and leads to

$$d\not{a} \int \widetilde{G}(A, B)\, \overline{\psi}_b(B)\, \psi_b(B)\, d\tau_B. \tag{35}$$

The mass of a is therefore given by

$$m(A) = \sum_b \int \widetilde{G}(A, B)\, \overline{\psi}_b(B)\, \psi_b(B)\, d\tau_B, \tag{36}$$

the integrals in (35), (36), being taken over all spacetime. This compares with

$$m(A) = \sum_b \int \widetilde{G}(A, B)\, db \tag{37}$$

in the classical theory. In place of the unique classical path of particle b in the line integral of (37), we now have a distribution of paths represented by the wave function ψ_b, which gives rise to a scalar contribution

$$\bar{\psi}_b(B)\,\psi_b(B)\,d\tau_B$$

from the four-dimensional volume element $d\tau_B$.

When the behavior of particle b can be discussed within the framework of the nonrelativistic approximation, the scalar $\bar{\psi}_b\psi_b$ is not much different from the time component of $\bar{\psi}_b\gamma^i\psi_b$, i.e., the probability density. To see this, use the Dirac representation

$$\gamma_4 = \begin{pmatrix} 1 & 0 \\ 0 & -1 \end{pmatrix}, \qquad \boldsymbol{\gamma} = \begin{pmatrix} 0 & \boldsymbol{\sigma} \\ -\boldsymbol{\sigma} & 0 \end{pmatrix}. \tag{38}$$

Writing

$$\psi_b = \begin{pmatrix} \psi_1 \\ \psi_2 \end{pmatrix}, \tag{39}$$

ψ_2 is then less than ψ_1 by the order of the velocity of the particle, and

$$\bar{\psi}_b\psi_b = \psi_b{}^*\gamma_4\psi_b \simeq \psi_1{}^*\psi_1, \tag{40}$$

which is also the large term in $\bar{\psi}_b\gamma_4\psi_b$. Since the nonrelativistic approximation can be used for most of the particles in the universe, the dominant contribution to the mass of a particle is therefore to be calculated from the probability densities of the particles. Instead of a unique classical path weighted by proper time increments db, we now have a four-dimensional distribution weighted essentially by the probability density.

Conceptually, we are not required to change our earlier classical picture of the origin of mass in any very major way. But what of the theory of gravitation? What happens if we attempt to generalize our ideas from Minkowski space to a Riemannian metric? These questions will be considered in the rest of this chapter.

§8.3. DIRAC PARTICLE IN A GRAVITATIONAL FIELD (LECTURE 20)

We write the Dirac equation in the form

$$\gamma^i\psi_{;i} + im\psi = 0, \tag{41}$$

taking the representation for γ^i such that the spinors making up ψ are separated:

$$\psi = \begin{pmatrix} u_\alpha \\ v^{\dot\beta} \end{pmatrix}. \tag{42}$$

In Minkowski space

$$\gamma^i = \begin{pmatrix} 0 & \sigma^i{}_{\alpha\dot\beta} \\ \sigma^{i\alpha\dot\beta} & 0 \end{pmatrix}, \tag{43}$$

with

$$\sigma^1{}_{\alpha\dot\beta} = \begin{pmatrix} 0 & 1 \\ 1 & 0 \end{pmatrix}, \qquad \sigma^2{}_{\alpha\dot\beta} = \begin{pmatrix} 0 & -i \\ i & 0 \end{pmatrix}, \qquad \sigma^3{}_{\alpha\dot\beta} = \begin{pmatrix} 1 & 0 \\ 0 & -1 \end{pmatrix}, \tag{44}$$

being the three Pauli matrices, and $\sigma^4{}_{\alpha\dot\beta}$ the unit 2×2 matrix. The spinor suffix $\alpha = 1, 2$, refers here to rows, and $\dot\beta = \dot 1, \dot 2$, refers to columns, whereas in $\sigma^{i\alpha\dot\beta}$ we take $\dot\beta$ to refer to rows and α to refer to columns. Contravariant and covariant spinors are related by

$$\begin{aligned} u^\alpha &= \varepsilon^{\alpha\beta} u_\beta, & u_\beta &= \varepsilon_{\alpha\beta} u^\alpha, \\ v^{\dot\alpha} &= \varepsilon^{\dot\alpha\dot\beta} v_{\dot\beta}, & v_{\dot\beta} &= \varepsilon_{\dot\alpha\dot\beta} v^{\dot\alpha}, \end{aligned} \tag{45}$$

where

$$\begin{aligned} &\varepsilon_{12} = \varepsilon^{12} = \varepsilon_{\dot 1\dot 2} = \varepsilon^{\dot 1\dot 2} = 1, \\ &\varepsilon_{21} = \varepsilon^{21} = \varepsilon_{\dot 2\dot 1} = \varepsilon^{\dot 2\dot 1} = -1, \\ &\varepsilon_{11} = \varepsilon_{22} = \varepsilon_{\dot 1\dot 1} = \varepsilon_{\dot 2\dot 2} = 0, \\ &\varepsilon^{11} = \varepsilon^{22} = \varepsilon^{\dot 1\dot 1} = \varepsilon^{\dot 2\dot 2} = 0. \end{aligned} \tag{46}$$

Using these rules one easily obtains

$$\sigma^{i\alpha\dot\beta} = -\sigma^i{}_{\alpha\dot\beta}, \qquad i = 1, 2, 3; \tag{47}$$

$$\sigma^{4\alpha\dot\beta} = \sigma^4{}_{\alpha\dot\beta} = \begin{pmatrix} 1 & 0 \\ 0 & 1 \end{pmatrix}. \tag{48}$$

The mixed spinor $\varepsilon^\alpha{}_\beta$ is given by

$$-\varepsilon^\alpha{}_\beta = \varepsilon_\beta{}^\alpha = \delta^\alpha{}_\beta, \tag{49}$$

where $\delta^\alpha{}_\beta$ is the Kronecker delta. It is not hard to see that

$$\sigma^{i\alpha\dot\beta}\sigma^k{}_{\alpha\dot\beta} = 2\eta^{ik}, \qquad \sigma^{i\alpha\dot\beta}\sigma_{i\gamma\dot\sigma} = 2\varepsilon^\alpha{}_\gamma\varepsilon^{\dot\beta}{}_{\dot\sigma}. \tag{50}$$

Still working in Minkowski space, we have

$$\psi_{;i} \equiv \frac{\partial\psi}{\partial x^i}. \tag{51}$$

After a Lorentz transformation, the same ψ would therefore continue to satisfy (41), provided the matrices γ^i transform like a contravariant vector, i.e., provided $\sigma^{i\alpha\dot\beta}$, $\sigma^i{}_{\alpha\dot\beta}$, transform like contravariant vectors. It is well-known that a spin transformation,

$$c_\gamma{}^\alpha \equiv \begin{bmatrix} c_1{}^1 & c_1{}^2 \\ c_2{}^1 & c_2{}^2 \end{bmatrix},$$

$$-c^\gamma{}_\alpha \equiv -\begin{bmatrix} c^1{}_1 & c^2{}_1 \\ c^1{}_2 & c^2{}_2 \end{bmatrix} = \begin{bmatrix} c_2{}^2 & -c_1{}^2 \\ -c_2{}^1 & c_1{}^1 \end{bmatrix}, \tag{52}$$

can be found such that a combination of a Lorentz transformation applied to the index i, of $c_\gamma{}^\alpha$ applied to the index α of $\sigma^i{}_{\alpha\dot\beta}$, of $-c^\gamma{}_\alpha$ applied to $\sigma^{i\alpha\dot\beta}$, and of the complex conjugate transformation applied to $\dot\beta$, has the net effect of leaving γ^i unchanged. From this important property, it follows that we can continue to use the same γ^i in (41) after a Lorentz transformation, provided an appropriate spin transformation is applied to the wave function ψ.

It is implicit here that the spin transformation (52) applies not only to ψ but to $\psi_{;i}$, because the coefficients defining a Lorentz transformation are constants, independent of spacetime position. The coefficients $c_\alpha{}^\gamma$ are therefore also independent of position, and can be taken outside the derivatives

$$\frac{\partial}{\partial x^i}(c_\gamma{}^\alpha u_\alpha) = c_\gamma{}^\alpha \frac{\partial u_\alpha}{\partial x^i}. \tag{53}$$

Next we consider how the above discussion is changed when a gravitational field is present. At any point we can choose a neighborhood in which the local metric can be taken to have the Minkowski form. With a suitable choice of coordinates, equations (41) to (53) hold in such a neighborhood. Let x^i be general coordinates applicable to all spacetime, with metric

$$ds^2 = g_{ik}\,dx^i\,dx^k. \tag{54}$$

In the neighborhood associated with a point X, make a coordinate transformation $x'^i = x'^i(x)$ such that

$$ds^2 = \eta_{ik}\, dx'^i\, dx'^k + O(ds^4). \tag{55}$$

From what has just been said, equations (41) to (53) can be taken to hold with respect to the x'^i coordinates in the neighborhood in question.

Consider now the inverse transformation $x^i = x^i(x')$. Letting the matrices σ^i transform like a contravariant vector, we get

$$g^{i\alpha\dot{\beta}} = \frac{\partial x^i}{\partial x'^j}\sigma^{j\alpha\dot{\beta}}, \qquad g^i{}_{\alpha\dot{\beta}} = \frac{\partial x^i}{\partial x'^j}\sigma^j{}_{\alpha\dot{\beta}}, \tag{56}$$

with respect to the x^i coordinates. Raising and lowering indexes, using g^{ik}, g_{ik}, in the usual way, we have

$$\begin{aligned} g^{i\alpha\dot{\beta}} g_{i\gamma\dot{\sigma}} &= g_{ik}\frac{\partial x^i}{\partial x'^j}\frac{\partial x^k}{\partial x'^l}\sigma^{j\alpha\dot{\beta}}\sigma^l{}_{\gamma\dot{\sigma}} \\ &= \eta_{jl}\sigma^{j\alpha\dot{\beta}}\sigma^l{}_{\gamma\dot{\sigma}} = 2\varepsilon^\alpha{}_\gamma\varepsilon^{\dot{\beta}}{}_{\dot{\sigma}}. \end{aligned} \tag{57}$$

Also

$$\begin{aligned} g^{i\alpha\dot{\beta}} g^k{}_{\alpha\dot{\beta}} &= \frac{\partial x^i}{\partial x'^j}\frac{\partial x^k}{\partial x'^l}\sigma^{j\alpha\dot{\beta}}\sigma^l{}_{\alpha\dot{\beta}} \\ &= 2\frac{\partial x^i}{\partial x'^j}\frac{\partial x^k}{\partial x'^l}\eta^{jl} = 2g^{ik}. \end{aligned} \tag{58}$$

Defining γ^i now by[1]

$$\gamma^i \equiv \begin{pmatrix} 0 & g^i{}_{\alpha\dot{\beta}} \\ g^{i\alpha\dot{\beta}} & 0 \end{pmatrix}, \tag{59}$$

what meaning must we attach to $\psi_{;i}$ in order that (41) may be taken to hold throughout the whole of spacetime? For the derivative of any spinor function of position u_α, write

$$u_{\alpha;i} = \frac{\partial u_\alpha}{\partial x^i} - L^\beta{}_{\alpha i} u_\beta, \tag{60}$$

where the coefficients $L^\beta{}_{\alpha i}$ of the affine connection transform as a covariant

---

[1] Once again we note that in writing $g^i{}_{\alpha\dot{\beta}}$ as a 2×2 matrix, α refers to rows and $\dot{\beta}$ to columns, whereas for $g^{i\alpha\dot{\beta}}$ the situation is reversed, with α specifying columns and $\dot{\beta}$ rows.

vector with respect to the index i. Denoting the complex conjugate of $L^{\beta}{}_{\alpha i}$ by $L^{\dot\beta}{}_{\dot\alpha i}$, the equation corresponding to (60) for any conjugate spinor $v_{\dot\beta}$ is

$$v_{\dot\beta;\,i} = \frac{\partial v_{\dot\beta}}{\partial x^i} - L^{\dot\alpha}{}_{\dot\beta i} v_{\dot\alpha}. \tag{61}$$

Consider a spin transformation $u_\alpha \to u'_\alpha = c_\alpha{}^\beta u_\beta$. Let $L'^{\beta}{}_{\alpha i}$ be the affine connection giving $u'_{\alpha;\,i}$,

$$u'_{\alpha;\,i} = \frac{\partial u'_\alpha}{\partial x^i} - L'^{\beta}{}_{\alpha i} u'_\beta. \tag{62}$$

Provided $L'^{\beta}{}_{\alpha i}$ satisfies

$$c_\beta{}^\gamma L'^{\beta}{}_{\alpha i} = c_\alpha{}^\beta L^\gamma{}_{\beta i} + \frac{\partial c_\alpha{}^\gamma}{\partial x^i}, \tag{63}$$

we have

$$u'_{\alpha;\,i} = c_\alpha{}^\beta u_{\beta;\,i}, \tag{64}$$

and the conjugate equation gives

$$v'_{\dot\beta;\,i} = c_{\dot\beta}{}^{\dot\gamma} v_{\dot\gamma;\,i}. \tag{65}$$

With this interpretation of spinor derivatives, and taking γ^i to transform like a contravariant vector, equation (41) is covariant with respect to both coordinate and spin transformations.

Using the unimodular property of the spin transformation (52), the choice

$$L^{\beta}{}_{\alpha i} = \frac{1}{4}\left(g^l{}_{\alpha\dot\lambda} g_m{}^{\beta\dot\lambda} \Gamma^m{}_{li} + g_l{}^{\beta\dot\lambda} \frac{\partial}{\partial x^i} g^l{}_{\alpha\dot\lambda}\right),$$

$$L^{\dot\beta}{}_{\dot\alpha i} = \frac{1}{4}\left(g^l{}_{\lambda\dot\alpha} g_m{}^{\lambda\dot\beta} \Gamma^m{}_{li} + g_l{}^{\lambda\dot\beta} \frac{\partial}{\partial x^i} g^l{}_{\lambda\dot\alpha}\right), \tag{66}$$

can be shown to satisfy (63). If we proceed to differentiate $g^i{}_{\alpha\dot\beta}$, using both the Christoffel symbols Γ and the affine connection L,

$$g^i{}_{\alpha\dot\beta;\,j} = \frac{\partial g^i{}_{\alpha\dot\beta}}{\partial x^j} + \Gamma^i{}_{jk} g^k{}_{\alpha\dot\beta} - L^\gamma{}_{\alpha j} g^i{}_{\gamma\dot\beta} - L^{\dot\sigma}{}_{\dot\beta j} g^i{}_{\alpha\dot\sigma}, \tag{67}$$

then it can be shown (see Appendix 6, p. 257) that the choice (66) also leads to

$$g^i{}_{\alpha\dot{\beta};j} \equiv 0. \tag{68}$$

This is analogous to the situation in Riemannian geometry, in which covariant differentiation is defined in such a way that $g_{ik;l} \equiv 0$.

The Dirac equation (41) for a particle in a gravitational field must be interpreted with respect to the more general form (59) for the γ^i matrices, and with respect to the differentiations (60), (61). We cannot proceed for general coordinate transformations in the same way as for Lorentz transformations. It is not possible to find a spin transformation $c_\alpha{}^\beta(X)$ which on being applied to $g^i{}_{\alpha\dot{\beta}}$ restores γ^i to the form (43) at all points X. To do so we would require

$$\sigma^i{}_{\alpha\dot{\beta}} = c_\alpha{}^\gamma c_{\dot{\beta}}{}^{\dot{\lambda}} g^i{}_{\gamma\dot{\lambda}} = c_\alpha{}^\gamma c_{\dot{\beta}}{}^{\dot{\lambda}} \sigma^j{}_{\gamma\dot{\lambda}} \frac{\partial x^i}{\partial x'^j}, \tag{69}$$

$x'^j = x'^j(x)$ being the coordinate transformation that reduces (59) locally to the form (43). It is not hard to show that (69) can be expressed in the form

$$\frac{\partial x'^k}{\partial x^i} = c_\alpha{}^\gamma c_{\dot{\lambda}}{}^{\dot{\beta}} \sigma_i{}^{\alpha\dot{\lambda}} \sigma^k{}_{\gamma\dot{\beta}}. \tag{70}$$

The matrix $c_\alpha{}^\gamma$ is unimodular,

$$\varepsilon_{\gamma\lambda} c_\alpha{}^\gamma c_\beta{}^\lambda = \varepsilon_{\alpha\beta}, \tag{71}$$

containing only six real parameters, whereas for a general coordinate transformation there are 16 independent quantities on the lefthand side of (70). Evidently, then, it is not possible to find a spin transformation that satisfies (70) generally. For the special case of Lorentz transformations in Minkowski space, there are only six independent quantities involved in the coefficients $\partial x'^k/\partial x^i$; so (70) can be satisfied for that case.

§8.4. THE GRAVITATIONAL FIELD OF A DIRAC PARTICLE

In the classical theory a particle is considered to follow a unique path. Associated with the path is an action S, which is used in two ways. Writing $\delta S = 0$ for a variation of the path with respect to a given gravitational field leads to equations determining the path. On the other hand, varying the geometry, $g_{ik} \to g_{ik} + \delta g_{ik}$, with respect to a given coordinate path leads to a nonzero variation of the action, which can be written in the form

$$\delta S = -\frac{1}{2}\int \delta g_{ik}\, T^{ik}{}_{(\mathrm{mat})}\sqrt{-g}\, d^4x, \tag{72}$$

where

$$T^{ik}{}_{(\mathrm{mat})}(X) = \int \frac{\delta_4(X, A)}{\sqrt{-g(A)}}\, m(A)\frac{da^i}{da}\frac{da^k}{da}\, da, \tag{73}$$

the notation being the same as in previous chapters. In this section we shall follow a similar procedure for the quantum case, not in terms of the path-integral method, but from a more empirical point of view, our aim being to obtain the quantum form of (73). In the classical case, (73) determines the effect of the particle on the gravitational field, in accordance with the Einstein equations [or with equations (117) of Chapter 6]. We seek now for the corresponding gravitational effect of a particle according to quantum mechanics.

In place of a unique classical path, we have a wave function ψ satisfying (41). We ask for the form of the invariant action S that would lead to (41), starting from the condition

$$\delta S = 0, \qquad \psi \to \psi + \delta\psi, \tag{74}$$

the metric tensor $g_{ik}(X)$ being considered a given function of position. The answer to this question is

$$S = \int (i\bar{\psi}\gamma^j\psi_{;j} - m\bar{\psi}\psi)\sqrt{-g}\, d^4x, \tag{75}$$

the condition (74) being applied to an arbitrary finite 4-volume V, with $\delta\psi$ set zero on the boundary of V. Here $\bar{\psi}$ is given by

$$\bar{\psi} = (v^\alpha, u_{\dot{\beta}}), \tag{76}$$

and

$$\bar{\psi}\gamma^j\psi_{;j} = g^{j\alpha\dot{\beta}}[u_{\dot{\beta}}u_{\alpha;j} + v_\alpha v_{\dot{\beta};j}], \tag{77}$$

$$\bar{\psi}\psi = u_\alpha v^\alpha + u_{\dot{\beta}}v^{\dot{\beta}}. \tag{78}$$

In order to verify that (75) does indeed lead to (41), we have to consider the variation δS resulting from

$$u_\alpha \to u_\alpha + \delta u_\alpha, \qquad v_{\dot\beta} \to v_{\dot\beta} + \delta v_{\dot\beta}.$$

Consider the particular term

$$i \int_V g^{j\alpha\dot\beta} u_{\dot\beta} (\delta u_\alpha)_{;j} \sqrt{-g}\, d^4x, \tag{79}$$

where

$$(\delta u_\alpha)_{;j} = \frac{\partial \delta u_\alpha}{\partial x^j} - L^\beta{}_{\alpha j}\, \delta u_\beta. \tag{80}$$

Integrating by parts, and using $\delta u_\alpha = 0$ on the boundary of V, we can write (79) in the form

$$-i \int_V \left[\frac{\partial}{\partial x^j} (g^{j\alpha\dot\beta} \sqrt{-g}\, u_{\dot\beta}) + g^{j\gamma\dot\beta} L^\alpha{}_{\gamma j} u_{\dot\beta} \sqrt{-g} \right] \delta u_\alpha\, d^4x. \tag{81}$$

The integrand of (81) is an invariant under coordinate transformations. At any point X choose coordinates such that $\partial g_{ik}/\partial x^l = 0$ for all values of i, k, l. Then the term in brackets in (81) can be written at X in the form

$$g^{j\alpha\dot\beta} \sqrt{-g} \frac{\partial u_{\dot\beta}}{\partial x^j} + u_{\dot\beta} \sqrt{-g} \left(\frac{\partial g^{j\alpha\dot\beta}}{\partial x^j} + g^{j\gamma\dot\beta} L^\alpha{}_{\gamma j} \right). \tag{82}$$

Since, moreover, our choice for $L^\alpha{}_{\gamma j}$ gives

$$g^{i\alpha\dot\beta}{}_{;j} = \frac{\partial g^{i\alpha\dot\beta}}{\partial x^j} + \Gamma^i{}_{jk} g^{k\alpha\dot\beta} + L^\alpha{}_{\gamma j} g^{i\gamma\dot\beta} + L^{\dot\beta}{}_{\dot\lambda j} g^{i\alpha\dot\lambda} \equiv 0,$$

this being the contravariant spin form of (68), and $\Gamma^i{}_{jk} = 0$ at X, the term in parentheses in (82) is simply $-L^{\dot\beta}{}_{\dot\lambda j} g^{j\alpha\dot\lambda}$, and (82) is just

$$g^{j\alpha\dot\beta} \sqrt{-g}\, u_{\dot\beta;j}. \tag{83}$$

Because (83) is of covariant form, the integral (81) is equal to

$$-i \int_V g^{j\alpha\dot\beta} \sqrt{-g}\, u_{\dot\beta;j}\, \delta u_\alpha\, d^4x \tag{84}$$

in any coordinate system. It is now easy to see that the variation of (75) is

$$\begin{aligned}\delta S = &\int \sqrt{-g}\,\delta u_\alpha(-ig^{j\alpha\dot\beta}u_{\dot\beta;j} - mv^\alpha)\,d^4x \\ &+ \int \sqrt{-g}\,\delta u_{\dot\beta}(ig^{j\alpha\dot\beta}u_{\alpha;j} - mv^{\dot\beta})\,d^4x \\ &+ \int \sqrt{-g}\,\delta v_\alpha(ig^{j\alpha\dot\beta}v_{\dot\beta;j} + mu^\alpha)\,d^4x \\ &+ \int \sqrt{-g}\,\delta v_{\dot\beta}(-ig^{j\alpha\dot\beta}v_{\alpha;j} + mu^{\dot\beta})\,d^4x. \end{aligned} \tag{85}$$

The first two terms of δS are complex conjugates, as are the last two terms. Hence we require the real part of the following expression to vanish:

$$\int \sqrt{-g}\,\delta u_\alpha(-ig^{j\alpha\dot\beta}u_{\dot\beta;j} - mv^\alpha)\,d^4x + \int \sqrt{-g}\,\delta v_\alpha(ig^{j\alpha\dot\beta}v_{\dot\beta;j} + mu^\alpha)\,d^4x. \tag{86}$$

With δu_α, δv_α, arbitrary, this requires

$$\begin{aligned} ig^{j\alpha\dot\beta}u_{\dot\beta;j} + mv^\alpha &= 0, \\ ig^{j\alpha\dot\beta}v_{\dot\beta;j} + mu^\alpha &= 0. \end{aligned} \tag{87}$$

Taking the complex conjugate of the first of these,

$$g^{j\alpha\dot\beta}u_{\alpha;j} + imv^{\dot\beta} = 0, \tag{88}$$

and raising and lowering indices in the second,

$$g^j{}_{\alpha\dot\beta}v^{\dot\beta}{}_{;j} + imu_\alpha = 0. \tag{89}$$

Equations (88), (89), combine to give (41), as we required.

A similar argument to that which led from (79) to (84) gives

$$i\int g^{j\alpha\dot\beta}(u_{\dot\beta}u_{\alpha;j} + v_\alpha v_{\dot\beta;j})\sqrt{-g}\,d^4x = \frac{1}{2}i\int g^{j\alpha\dot\beta}(u_{\dot\beta}u_{\alpha;j} - u_{\dot\beta;j}u_\alpha + v_\alpha v_{\dot\beta;j} - v_{\alpha;j}v_{\dot\beta})\sqrt{-g}\,d^4x. \tag{90}$$

Hence (72) gives

$$\frac{1}{2} i\, \delta \int g^{j\alpha\dot{\beta}}(u_{\dot{\beta}} u_{\alpha;j} - u_{\dot{\beta};j} u_\alpha + v_\alpha v_{\dot{\beta};j} - v_{\alpha;j} v_{\dot{\beta}}) \sqrt{-g}\, d^4x$$

$$- m \int (u_\alpha v^\alpha + u_{\dot{\beta}} v^{\dot{\beta}})\, \delta \sqrt{-g}\, d^4x$$

$$= -\frac{1}{2} \int \delta g_{ik} T^{ik}{}_{(\text{mat})} \sqrt{-g}\, d^4x = \frac{1}{2} \int \delta g^{ik} T_{(\text{mat})ik} \sqrt{-g}\, d^4x, \quad (91)$$

our aim being now to calculate $T_{(\text{mat})ik}$. It is to be noticed that the variation of $\sqrt{-g}$ on the lefthand side of (91) makes no contribution, because of (41).

From (56) it follows that $g^{i\alpha\dot{\beta}}$ and $g^{i\dot{\beta}\alpha}$ must be complex conjugates. In spite of this restriction, a variation $g^{i\alpha\dot{\beta}} \to g^{i\alpha\dot{\beta}} + \delta g^{i\alpha\dot{\beta}}$ contains 16 independent parameters, compared to the ten independent variations contained in g_{ik}. This can be seen explicitly by considering the choice

$$\delta g^{i\alpha\dot{\beta}} = \frac{1}{2} g_k{}^{\alpha\dot{\beta}}\, \delta g^{ik} + \theta^\alpha{}_\gamma g^{i\gamma\dot{\beta}} + \theta^{\dot{\beta}}{}_{\dot{\lambda}} g^{i\alpha\dot{\lambda}}, \quad (92)$$

where $\theta^\alpha{}_\gamma + \delta^\alpha{}_\gamma$ is any infinitesimal unimodular transformation satisfying $\theta^\alpha{}_\alpha = 0$. The choice (92) meets the requirement that $\delta g^{i\alpha\dot{\beta}}$ and $\delta g^{i\dot{\beta}\alpha}$ must be complex conjugates. From (58)

$$2\, \delta g^{ik} = \delta g^{i\alpha\dot{\beta}} g^k{}_{\alpha\dot{\beta}} + \delta g^k{}_{\alpha\dot{\beta}} g^{i\alpha\dot{\beta}}. \quad (93)$$

Raising and lowering indices in the second term on the righthand side of (93), we have

$$\delta g^{ik} = \delta g^{ik} + g^{ik}(\theta^\alpha{}_\alpha + \theta^{\dot{\beta}}{}_{\dot{\beta}}) = \delta g^{ik}. \quad (94)$$

The part of (92) involving the infinitesimal transformation on the spinor indices thus has no effect on the metric tensor. Nor has such a spin transformation any effect on S. It follows that if we start with variations $\delta g^{i\alpha\dot{\beta}}$ of the form (92), components of $\theta^\alpha{}_\beta$ being nonzero, it is possible to apply an inverse transformation $-\theta^\alpha{}_\beta + \delta^\alpha{}_\beta$ to the spinor indices in S. After such an inverse transformation, $\delta g^{i\alpha\dot{\beta}}$ is given by

$$\delta g^{i\alpha\dot{\beta}} = \frac{1}{2} g_k{}^{\alpha\dot{\beta}}\, \delta g^{ik}, \quad (95)$$

and this is without S being changed.

We shall now calculate the variation δS resulting from (95), noting that the variations

$$u_\alpha \to u_\alpha + \delta u_\alpha,\ v_{\dot\beta} \to v_{\dot\beta} + \delta v_{\dot\beta},$$

resulting from the transformation $-\theta^\alpha{}_\beta + \delta^\alpha{}_\beta$, and its complex conjugate, do not change S. In Appendix 7 (p. 259) we show that $\delta u_{\alpha;\,j}$, $\delta v_{\dot\beta;\,j}$ also do not change S. Accordingly, we have

$$\delta S = \frac{1}{2} i \int \delta g^{j\alpha\dot\beta}(u_{\dot\beta}u_{\alpha;\,j} - u_{\dot\beta;\,j}u_\alpha + v_\alpha v_{\dot\beta;\,j} - v_{\alpha;\,j}v_{\dot\beta})\sqrt{-g}\,d^4x. \quad (96)$$

Substituting from (95) for $\delta g^{j\alpha\dot\beta}$ gives

$$\delta S = \frac{1}{4} i \int \delta g^{jk} g_k{}^{\alpha\dot\beta}(u_{\dot\beta}u_{\alpha;\,j} - u_{\dot\beta;\,j}u_\alpha + v_\alpha v_{\dot\beta;\,j} - v_{\alpha;\,j}v_{\dot\beta})\sqrt{-g}\,d^4x$$

$$= \frac{1}{2}\int \delta g^{jk} T_{(\mathrm{mat})jk}\sqrt{-g}\,d^4x, \quad (97)$$

from which we have

$$4T_{(\mathrm{mat})jk} = ig_k{}^{\alpha\dot\beta}(u_{\dot\beta}u_{\alpha;\,j} - u_{\dot\beta;\,j}u_\alpha + v_\alpha v_{\dot\beta;\,j} - v_{\alpha;\,j}v_{\dot\beta})$$
$$+\, ig_j{}^{\alpha\dot\beta}(u_{\dot\beta}u_{\alpha;\,k} - u_{\dot\beta;\,k}u_\alpha + v_\alpha v_{\dot\beta;\,k} - v_{\alpha;\,k}v_{\dot\beta}). \quad (98)$$

In terms of ψ, $\bar\psi$,

$$T_{(\mathrm{mat})jk} = \frac{1}{4} i[\bar\psi\gamma_k\psi_{;j} + \bar\psi\gamma_j\psi_{;k} - \bar\psi_{;j}\gamma_k\psi - \bar\psi_{;k}\gamma_j\psi], \quad (99)$$

which is the required form of $T_{(\mathrm{mat})jk}$.

For a free particle of momentum p^i in Minkowski space,

$$T_{(\mathrm{mat})jk} = \frac{\psi^*\psi}{E}p_j p_k, \quad (100)$$

and setting $\psi^*\psi = n$ for a beam of such particles of proper density n, we have

$$T_{(\mathrm{mat})jk} = \frac{n}{E}p_j p_k. \quad (101)$$

In the nonrelativistic case $E \simeq m$, and (101) has the same form as a classical fluid of mass density mn. This close analogy with the classical case suggests that the gravitational effect of a Dirac particle is to be calculated by taking the energy-momentum tensor in the Einstein equation to be of the form (99).

The corresponding classical energy-momentum tensor of a particle a is

$$\int \frac{\delta_4(X, A)}{\sqrt{-g(A)}} m(A) \frac{da_j}{da} \frac{da_k}{da} \, da. \tag{102}$$

This differs from (99) in the critical respect that (102) is singular at the world line of the particle. If one were to solve the Einstein equations using (102) literally, instead of a smoothed approximation, the resulting metric would be singular at the particle. This difficulty is avoided by (99), which indeed can be regarded as a smoothed form of (102). This smoothing effect of quantum mechanics is critical to the concept of a "test particle"—that is to say, a particle which is considered to respond to the gravitational field without itself having a significant effect on the field. Smoothing on a scale much greater than the gravitational radius of the particle, $2Gm$, is needed for such a concept to be feasible. Unless the kinetic energy of a quantum particle is much larger than m, the wave function ψ smooths (99) on a scale greater than m^{-1}. Thus, as an order-of-magnitude statement, we require

$$\sim m^{-1} \gg 2Gm. \tag{103}$$

This condition is satisfied for actual particles by the very large factor $\sim 10^{40}$, a factor whose significance we shall consider in greater detail in the next chapter.

§8.5. THE GRAVITATIONAL EQUATIONS (LECTURE 21)

The discussion in the last two sections, although giving interesting results, is not logically satisfactory, since it associates a sophisticated quantum theory of particles with only a classical theory of gravitation. It would be best, of course, to lift our consideration of gravitation to a level of sophistication like that for particles, but such a program runs into a difficulty which we shall discuss in the next section. For the moment, then, we shall content ourselves with the opposite procedure, of degrading the theory of particles down to a level where it matches the classical theory of gravitation. This procedure has two advantages: first, it shows how the phenomenon of gravitation arises from a quantum point of view; and, second, it suggests an approach to a more refined consideration of the problem.

The relevant relationship between particles and the classical theory of gravitation is contained in the quasiclassical quantum approximation. The nonrelativistic Schrödinger equation for a particle of mass m in a potential $U(X)$ at point $X(\boldsymbol{x}, t)$ is

$$i\frac{\partial\psi}{\partial t} = -\frac{1}{2m}\left(\frac{\partial^2\psi}{\partial x^2} + \frac{\partial^2\psi}{\partial y^2} + \frac{\partial^2\psi}{\partial z^2}\right) + U\psi. \tag{104}$$

It is always possible to choose real functions $A(X)$, $S(X)$, such that

$$\psi = A\exp(iS). \tag{105}$$

Thus writing $\psi = \psi_1 + i\psi_2$, where ψ_1, ψ_2, are real functions of $\boldsymbol{x}$, t, we take $\tan S = \psi_2/\psi_1, A\cos S = \psi_1$. Substituting (105) in (104), and separating real and imaginary parts, we get

$$\frac{\partial S}{\partial t} + \frac{1}{2m}(\boldsymbol{\nabla}S)^2 + U - \frac{1}{2mA}\left(\frac{\partial^2 A}{\partial x^2} + \frac{\partial^2 A}{\partial y^2} + \frac{\partial^2 A}{\partial z^2}\right) = 0, \tag{106}$$

$$\frac{\partial A}{\partial t} + \frac{A}{2m}\left(\frac{\partial^2 S}{\partial x^2} + \frac{\partial^2 S}{\partial y^2} + \frac{\partial^2 S}{\partial z^2}\right) + \frac{1}{m}\boldsymbol{\nabla}S\cdot\boldsymbol{\nabla}A = 0. \tag{107}$$

In the quasiclassical approximation, the function $A(X)$ is considered to be slowly varying with respect to the point X, so that the terms in (106) involving second derivatives of A are small. Neglecting these terms, (106) becomes the Hamilton-Jacobi equation of classical mechanics

$$\frac{\partial S}{\partial t} + \frac{1}{2m}(\boldsymbol{\nabla}S)^2 + U = 0. \tag{108}$$

And multiplying (107) by $2A$ leads immediately to

$$\frac{\partial A^2}{\partial t} + \operatorname{div}\left[\frac{A^2}{m}\boldsymbol{\nabla}S\right] = 0. \tag{109}$$

To interpret these equations, consider an initial time, t_0, say. At any point $\boldsymbol{x}$ where A^2 is nonzero, define a velocity by

$$\boldsymbol{v} = \frac{1}{m}\boldsymbol{\nabla}S. \tag{110}$$

With (110) as initial condition at $\boldsymbol{x}$, t_0, consider the classical trajectory. At any time t such a trajectory defines a velocity which continues to satisfy (110), since (110) applies at all points of a classical trajectory—provided of course that S satisfies (108). By considering trajectories for all points $\boldsymbol{x}$, t_0, at which $A^2 \neq 0$, we obtain a velocity field satisfying (110), and in terms of this velocity field

$$\frac{\partial A^2}{\partial t} + \operatorname{div}(A^2 \boldsymbol{v}) = 0, \tag{111}$$

which is like the equation of continuity of a fluid. The classical trajectories are thus the stream lines of a "probability fluid."

All these results hold more generally than for the wave equation (104), being applicable to any quantum system with a classical analogue (Dirac, 1947).

Denote a particular streamline by $\boldsymbol{a}(t)$, and for any point $\boldsymbol{x}$, t, define $\boldsymbol{\xi}(t)$ by

$$\boldsymbol{\xi}(t) = \boldsymbol{x} - \boldsymbol{a}. \tag{112}$$

We have

$$S(\boldsymbol{x}, t) = S[\boldsymbol{a}(t)] + [\boldsymbol{\xi} \cdot \boldsymbol{\nabla} S]_{\boldsymbol{a}, t} + \cdots. \tag{113}$$

Here $S[\boldsymbol{a}(t)]$ is the classical action evaluated for the path $\boldsymbol{a}(t)$. When higher-order terms than the first in $\boldsymbol{\xi}$ are neglected, the phase factor $\exp(iS)$ is thus

$$\begin{aligned} \exp iS(\boldsymbol{x}, t) &= \exp iS[\boldsymbol{a}(t)] \cdot \exp i(\boldsymbol{\xi} \cdot \boldsymbol{\nabla} S) \\ &= \exp iS[\boldsymbol{a}(t)] \cdot \exp i(m\dot{\boldsymbol{a}} \cdot \boldsymbol{\xi}), \end{aligned} \tag{114}$$

and

$$\psi = \exp iS[\boldsymbol{a}(t)] \cdot \exp i(m\dot{\boldsymbol{a}} \cdot \boldsymbol{\xi}) \cdot A. \tag{115}$$

To this approximation, the wave function consists of the probability factor $A(\boldsymbol{x}, t)$, with A^2 moving like a classical fluid, an oscillating factor $\exp i(m\dot{\boldsymbol{a}} \cdot \boldsymbol{\xi})$ corresponding to momentum $m\dot{\boldsymbol{a}}$, and a classical phase $S[\boldsymbol{a}(t)]$, which at time t is independent of $\boldsymbol{x}$. This last factor plays no role in quantum calculations, since the wave function is considered at any time t to be arbitrary to within a constant phase factor. Yet it is this last factor which plays a crucial role in determining the classical theory of gravitation.

In order to deal with a many-particle system, we refer to (115) as the wave function for particle a, and denote it by ψ_a. For particle b we have correspondingly

$$\psi_b(\boldsymbol{x}_b, t) = \exp iS[\boldsymbol{b}(t)] \cdot \exp i(m\dot{\boldsymbol{b}} \cdot \boldsymbol{\eta}) \cdot B, \tag{116}$$

where $\boldsymbol{\eta}(t) = \boldsymbol{x}_b - \boldsymbol{b}(t)$. For particles $a, b, \ldots$ that are well-separated from each other, the total wave function $\underline{\psi}$ is given by the simple product

$$\underline{\psi}(\boldsymbol{x}_a, \boldsymbol{x}_b, \ldots, t) = \psi_a(\boldsymbol{x}_a, t)\psi_b(\boldsymbol{x}_b, t)\ldots. \tag{117}$$

Appearing now in $\underline{\psi}$ is the factor

$$\exp\left\{i \sum_a S[\boldsymbol{a}(t)]\right\}, \tag{118}$$

in which $S[\boldsymbol{a}(t)]$, $S[\boldsymbol{b}(t)], \ldots$, are the action values for the classical paths $\boldsymbol{a}(t), \boldsymbol{b}(t), \ldots$. Because we are working from the nonrelativistic Schrödinger equation (104), $S[\boldsymbol{a}(t)]\ldots$ would in the first instance be calculated nonrelativistically. However, it is not necessary to go to fully relativistic quantum mechanics in order to see that a relativistic calculation,

$$S[\boldsymbol{a}(t)] = -m \int da, \ldots, \tag{119}$$

gives a better approximation for the free-particle case than

$$\frac{1}{2} m \int \dot{\boldsymbol{a}}^2 \, dt. \tag{120}$$

In the nonrelativistic approximation to

$$K_+(2; 1) = (\not{\nabla}_2 - im)\left[\frac{1}{4\pi}\,\delta(s_{21}^2) - \frac{m}{8\pi s_{21}}\, H_1^{(2)}(m s_{21})\right], \tag{121}$$

we have

$$T = t_2 - t_1 \gg |\boldsymbol{x}_2 - \boldsymbol{x}_1| = |\boldsymbol{X}| \text{ say}, \tag{122}$$

$$m s_{21} \simeq m\left(T - \frac{X^2}{2T}\right) \gg 1, \tag{123}$$

giving

$$\delta({s^2}_{21}) = 0, \qquad H_1^{(2)}(m s_{21}) \simeq \left(\frac{2}{\pi m s_{21}}\right)^{1/2} \exp\left\{-i\left(m s_{21} - \frac{3\pi}{4}\right)\right\}. \tag{124}$$

In terms of the Dirac representation

$$\gamma_4 = \begin{pmatrix} 1 & 0 \\ 0 & -1 \end{pmatrix}, \quad \boldsymbol{\gamma} = \begin{pmatrix} 0 & \boldsymbol{\sigma} \\ -\boldsymbol{\sigma} & 0 \end{pmatrix}, \tag{125}$$

we get

$$K_+(2;1) \simeq \begin{bmatrix} 1 & -\dfrac{\boldsymbol{\sigma}\cdot\boldsymbol{X}}{2T} \\ \dfrac{\boldsymbol{\sigma}\cdot\boldsymbol{X}}{2T} & 0 \end{bmatrix} \left(\frac{m}{2\pi i T}\right)^{3/2} \exp(-ims_{21}). \qquad (126)$$

Using (123), and remembering that in the Dirac representation it is the unity 2×2 matrix in the upper lefthand corner of (126) that propagates the large part of the wave function, we obtain the free-particle propagator of the Schrödinger theory,

$$\left(\frac{m}{2\pi i T}\right)^{3/2} \exp\left(\frac{imX^2}{2T} - imT\right). \qquad (127)$$

There is no need, however, to degrade the $\exp(-ims_{21})$ factor in (126) to the form given in (127), and $\exp(-ims_{21})$ is just the factor implied by (118), (119), for any one of the particles a, b, ... Thus we can regard the factor (118) as being computed relativistically with respect to the paths $\boldsymbol{a}(t)$, $\boldsymbol{b}(t)$, ...

Our quantum discussion is in Minkowski space, where the classical paths of free particles are straight lines, in which case (119) gives the four-dimensional distance between the endpoints of the integral, thereby yielding a phase factor in (118) of the same form as $\exp(-ims_{21})$.

When a gravitational field is present, we transform locally to Minkowski space, calculating the phase factor of a particle in the manner described above. Transforming back to the original coordinate system does not change the wave function—we saw in §8.3 that for general coordinate transformations γ^i is a contravariant vector, whereas ψ is unchanged. Consequently the phase factor $\exp\{iS[\boldsymbol{a}(t)]\}$ is present in a general coordinate system. Since the determination of S according to (119) is invariant, we can indeed calculate $\exp\{iS[\boldsymbol{a}(t)]\}$ without making the transformation to Minkowski space, and the total phase factor (118) for all the particles can be similarly calculated.

Suppose we now admit a small change $g_{ik} \to g_{ik} + \delta g_{ik}$. All three factors in the wave function (115) of a particle will be altered. Variation of A is less important than variation of S because the latter is a phase factor, so that for $S \gg 1$ only a small fractional change $\delta S/S$ is in general needed to make a large change in ψ. And for particle systems considered over a time-interval very large compared to m^{-1}, the function S is indeed $\gg 1$. Furthermore, the first factor of (115) is much more sensitive to a change of S than the second factor, because $|\boldsymbol{\xi}|$ is small compared to the integral in (119). We see, therefore, that the factor (118) for a many-particle system, although playing no role in a purely quantum-mechanical problem, becomes the sensitive part of the wave function as soon as we consider a small change of geometry.

To formulate an explicit problem, consider the wave function $\underline{\psi}$ on a spacelike surface σ. The argument of (118) is

$$-\sum_a \int^\sigma m(A)\, da, \tag{128}$$

where the line integrals are taken up to σ, and where we now consider the possibility that the mass may depend on position. Indeed, if we interpret m in the same way as in earlier chapters, (128) would become

$$-\sum_a \sum_b \int^\sigma \int \tilde{G}(A, B)\, da\, db, \tag{129}$$

the integral with respect to db being unrestricted. It is not hard to see that (129) can be written in the form

$$-\frac{1}{2}\sum_a \sum_b \int\int \tilde{G}(A, B)\, da\, db - \frac{1}{2}\sum_a \sum_b \left[\int^\sigma \int^\sigma \tilde{G}(A, B)\, da\, db - \int_\sigma \int_\sigma \tilde{G}(A, B)\, da\, db\right], \tag{130}$$

and that for variations of geometry at σ there is no change in the second term of (130). The change of the first term is just the problem discussed in Chapter 6.

We now take the view that to investigate any physical problem in the neighborhood of σ, we must consider not a unique geometry at σ but all geometries. Thus the wave function $\underline{\psi}$ considered above is to be taken as the amplitude for a particular geometry. We have to add amplitudes for all geometries, which involves the summation

$$\sum_{\text{geometries}} \exp\left[-\frac{i}{2}\sum_a \sum_b \int\int \tilde{G}(A, B)\, da\, db\right]. \tag{131}$$

Only geometries close to that for which

$$\delta \sum_a \sum_b \int\int \widetilde{G}(A, B)\, da\, db = 0, \qquad g_{ik} \to g_{ik} + \delta g_{ik}, \qquad (132)$$

will be recognized in this summation. Thus a physical problem considered at a particular location (neighborhood of σ) picks out a geometry g_{ik} determined by applying (132) at that location.

We found in Chapter 6 that (132) leads in a certain conformal frame to the Einstein equations. We now see that a solution $g_{ik}(X)$ of the Einstein equations is not to be interpreted as determining a unique geometry for the world; $g_{ik}(X)$ in such a case is the form of metric tensor which dominates a physical problem at X, but all other geometries, not satisfying the Einstein equations, are to be considered logically possible.

Returning to (106), what is the condition for the terms in A to be small? Let the probability A^2 be distributed approximately uniformly through a spatial region of dimensions $\sim L$. Then the terms in A are of order $m^{-1}L^{-2}$, whereas for a nonrelativistic particle $\partial S/\partial t$ is of order m and $(\nabla S)^2/2m$ is of order $m\mathbf{v}^2$. Thus the required condition is $L \gg m^{-1}|\mathbf{v}|^{-1}$.

§8.6. QUANTUM GRAVITATION (LECTURE 22)

It is possible to define quantum gravitation in close analogy to quantum mechanics. We pass from the unique path of classical theory to quantum mechanics by summing amplitudes for all paths. Similarly, we go from the unique geometry of classical gravitation to quantum gravitation by summing amplitudes for all geometries.

We could seek to give expression to this idea by generalizing the concept of a path integral to include all geometries as well as all paths. However, from the outset we encounter the difficulty that such a procedure contains redundancies. We cannot vary the ten components of $g_{ik}(X)$ for each point X and the paths of the particles independently. To see this, make a small change $a^i(\tau) \to a^i(\tau) + \delta a^i(\tau)$ in the parametric representation of a path a. Now choose an infinitesimal coordinate transformation $x'^i = x'^i(x)$ such that $a'^i(\tau) = a^i(\tau) - \delta a^i(\tau)$ for all values of the parameter τ. That is to say, in the x' coordinate frame the world line is represented by the same functions of τ as before. The coordinate transformation changes $g_{ik}(X)$ to $g_{ik}(X) + \delta g_{ik}(X)$, say. A change of geometry $g_{ik}(X) \to g_{ik}(X) - \delta g_{ik}(X)$ then leaves g_{ik} the same as before, and we are back to the original situation. Of the ten degrees of freedom of the $g_{ik}(X)$ at each point X, six belong to the geometry itself, and four give freedom in the labelling of coordinates.

It follows that if we use a path integral containing a summation with respect to $g_{ik}(X)$, as well as with respect to the paths of particles, we shall be obliged to restrict the integral in some way in order that the appropriate degrees of freedom do not appear more than once. The alternative is to formulate the geometry of spacetime in a way that has only six independent functions of position, instead of the ten functions of the metric tensor. This implies a departure from the usual way of describing Riemannian geometry. Hitherto, the quantization of gravitation has always been discussed within the framework of the first alternative. Here we shall follow the second. We shall not carry the discussion beyond the objective just stated, however—further discussion of quantum gravitation would go beyond the scope of this book. Nevertheless, we regard the following discussion as a necessary prerequisite to an effective quantum theory of gravitation, particularly for the strong-field case.

A Subgroup of U_4

By using the quantities $g^i{}_{\alpha\dot\beta}$, discussed earlier in this chapter, any vector A_i in Riemannian space can be expressed as a spinor

$$A_i g^i{}_{\alpha\dot\beta} = A_{\alpha\dot\beta}, \text{ say.} \tag{133}$$

Given $A_{\alpha\dot\beta}$ it is easy to recover A_i. Thus

$$g_i{}^{\alpha\dot\beta} A_{\alpha\dot\beta} = g_i{}^{\alpha\dot\beta} g^k{}_{\alpha\dot\beta} A_k = 2A_i, \tag{134}$$

since $g_i{}^{\alpha\dot\beta} g^k{}_{\alpha\dot\beta} = 2g_i{}^k$. Evidently, then, $A_{\alpha\dot\beta}$ is an equivalent way of specifying A_i.

Physical vectors in Riemannian space can always be expressed in a particular spinor form. Writing $A_{\alpha\dot\beta}$ in the column form

$$A_L = \begin{bmatrix} A_{1\dot1} \\ A_{2\dot2} \\ A_{1\dot2} \\ A_{2\dot1} \end{bmatrix}; \qquad L = 1, 2, 3, 4, \tag{135}$$

then A_1, A_2, are real, and A_3, A_4, are complex conjugates. This follows because the $g^i{}_{\alpha\dot\beta}$ are Hermitian.

Consider transformations of A_L through unitary matrices

$$A_M' = U_{ML} A_L, \tag{136}$$

U_{ML} being the components of a unitary matrix U. We ask for A_M' to have the same "reality property" as A_L, namely A_1', A_2', real and A_3', A_4', complex conjugates. It will turn out that the matrices U which preserve this property, and which hence correspond to real transformations in Riemannian space, form a six-parameter subgroup of the four-dimensional unitary group U_4.

Taking, first, the case in which U is infinitesimal, we can write

$$U = 1 + i\varepsilon H, \tag{137}$$

where ε is infinitesimal and H is Hermitian. The most general form of H is

$$\begin{bmatrix} a & b & c & d \\ \bar{b} & e & f & g \\ \bar{c} & \bar{f} & h & k \\ \bar{d} & \bar{g} & \bar{k} & l \end{bmatrix} \tag{138}$$

in which a, e, h, l are real. Writing

$$A_L = \begin{bmatrix} p \\ q \\ r \\ \bar{r} \end{bmatrix} \tag{139}$$

with p, q, real, (136), (137), (138), and (139) give

$$A_M' = A_M + i\varepsilon \begin{bmatrix} ap & + & bq & + & cr & + & d\bar{r} \\ \bar{b}p & + & eq & + & fr & + & g\bar{r} \\ \bar{c}p & + & \bar{f}q & + & hr & + & k\bar{r} \\ \bar{d}p & + & \bar{g}q & + & \bar{k}r & + & l\bar{r} \end{bmatrix}. \tag{140}$$

Since p, q, r, are arbitrary, the requirement that A_1', A_2', be real and A_3', A_4', be complex conjugates gives

$$\begin{gathered} a = 0, \quad b \text{ imaginary}, \quad \bar{c} = -d, \\ e = 0, \quad \bar{f} = -g, \quad h = -l, \quad k = 0. \end{gathered} \tag{141}$$

In terms of real α, β, γ, δ, λ, μ, we can therefore express H in the form

$$H = \begin{bmatrix} 0 & i\alpha & \gamma + i\delta & -\gamma + i\delta \\ -i\alpha & 0 & \lambda + i\mu & -\lambda + i\mu \\ \gamma - i\delta & \lambda - i\mu & \beta & 0 \\ -\gamma - i\delta & -\lambda - i\mu & 0 & -\beta \end{bmatrix}. \tag{142}$$

That is to say,

$$H = \alpha I_1 + \beta I_2 + \gamma I_3 + \delta I_4 + \lambda I_5 + \mu I_6, \tag{143}$$

where

$$I_1 = \begin{bmatrix} 0 & i & 0 & 0 \\ -i & 0 & 0 & 0 \\ 0 & 0 & 0 & 0 \\ 0 & 0 & 0 & 0 \end{bmatrix}, \quad I_2 = \begin{bmatrix} 0 & 0 & 0 & 0 \\ 0 & 0 & 0 & 0 \\ 0 & 0 & 1 & 0 \\ 0 & 0 & 0 & -1 \end{bmatrix},$$

$$I_3 = \begin{bmatrix} 0 & 0 & 1 & -1 \\ 0 & 0 & 0 & 0 \\ 1 & 0 & 0 & 0 \\ -1 & 0 & 0 & 0 \end{bmatrix}, \quad I_4 = \begin{bmatrix} 0 & 0 & i & i \\ 0 & 0 & 0 & 0 \\ -i & 0 & 0 & 0 \\ -i & 0 & 0 & 0 \end{bmatrix},$$

$$I_5 = \begin{bmatrix} 0 & 0 & 0 & 0 \\ 0 & 0 & 1 & -1 \\ 0 & 1 & 0 & 0 \\ 0 & -1 & 0 & 0 \end{bmatrix}, \quad I_6 = \begin{bmatrix} 0 & 0 & 0 & 0 \\ 0 & 0 & i & i \\ 0 & -i & 0 & 0 \\ 0 & -i & 0 & 0 \end{bmatrix}, \tag{144}$$

are all Hermitian.

We show now that all unitary matrices, whether infinitesimal or not, with the required property for A_M' are determined by Hermitian matrices of the form of (143). To demonstrate this result, we make use of the algebraic property that a unitary matrix U gives rise to a Hermitian matrix H through the relation

$$iH = (1 - U)(1 + U)^{-1}. \tag{145}$$

The meaning of (145) is that iH when multiplied by $1 + U$ gives $1 - U$. Expanding, we obtain

$$\begin{aligned} iH &= (1 - U)(1 - U + U^2 - U^3 \ldots) \\ &= 1 - 2U + 2U^2 - 2U^3 \ldots, \end{aligned} \tag{146}$$

from which we see that if U applied to a real vector A_L yields a real vector

A_M' then iH applied to A_L also yields a real vector. Hence H must be of the form (143). Since (145) can be inverted to give

$$U = (1 - iH)(1 + iH)^{-1}, \tag{147}$$

the result that U can be generated by a Hermitian matrix of the form (143) is proven. Moreover a matrix H of the form (143) yields a matrix U, by (147), with the required reality property.

It is also not difficult to show that the subset of unitary matrices with the reality property in question form a group. If U_1, U_2, operating separately on A_L each yield a real vector, then so does the product U_2U_1. The unity matrix obviously has the required property. If U has the required property, then so has iH given by (145). Now

$$U^{-1} = (1 + iH)(i - iH)^{-1} \tag{148}$$

can be expanded in powers of iH, from which we see that U^{-1} also has the required property. The unitary matrices in question therefore form a group, which is clearly a subgroup of U_4. This subgroup is not normal (invariant) and so cannot be used to factor U_4.

There is a singular case which must be mentioned. It is necessary in using (145) that U shall not have -1 as an eigenvalue. This case can be covered by a limiting argument applied to the parameters α, β, γ, δ, λ, μ, in (143).

§8.7. A NEW FORMULATION OF RIEMANNIAN GEOMETRY

Let α, β, γ, δ, λ, μ, be single-valued, well-behaved functions of four real parameters x^i; $i = 1, 2, 3, 4$. We shall refer to these parameters as "coordinates." Denote an ordered chain of infinitesimal hermitian spinors, each with an undotted and a dotted index, by $a^{(n)}{}_{\alpha\dot{\beta}}$; $n = 1, 2, \ldots$. Consider the following relations:

$$x^i{}_{(n)} = \sum_{r=1}^{n} g_{(r)}{}^{i\alpha\dot{\beta}} a^{(r)}{}_{\alpha\dot{\beta}}, \tag{148}$$

$$g_{(n+1)}{}^{i\alpha\dot{\beta}} = U[x_{(n)}]\sigma^{i\alpha\dot{\beta}}. \tag{149}$$

Here $U[x_{(n)}]$ is the unitary matrix given by (143), (147), with α, β, γ, δ, λ, μ, taken at the parametric values $x^i{}_{(n)}$. For a given chain $a^{(n)}{}_{\alpha\dot{\beta}}$, we can solve equations (148), (149), iteratively, beginning with $x_{(0)}{}^i = 0$, $U[x_{(0)}] = 1$. By the spinors being infinitesimal, we mean that each term in the series (148) is infinitesimal.

At each step of such a spinor chain, we thus have four coordinate values x^i, starting with $x^i = 0$ ($i = 1, 2, 3, 4$) at the beginning of the chain. Geometrically, we can think of a spinor chain as representing a path in coordinate space, beginning at the origin of coordinates. The same coordinate values x^i can be arrived at through different spinor chains. Thus we can have

$$\sum_{r=1}^{n} g_{(r)}{}^{i\alpha\dot{\beta}} a^{(r)}{}_{\alpha\dot{\beta}} = \sum_{r=1}^{n'} g_{(r)}{}^{i\alpha\dot{\beta}} a'^{(r)}{}_{\alpha\dot{\beta}} \tag{150}$$

for two distinct chains $a^{(r)}{}_{\alpha\dot{\beta}}$, $a'^{(r)}{}_{\alpha\dot{\beta}}$; $r = 1, 2, \ldots$, the identity of coordinate values occurring from n, n', steps, respectively. The coordinate values associated with $r < n$ for $a^{(r)}{}_{\alpha\dot{\beta}}$, and $r < n'$ for $a'^{(r)}{}_{\alpha\dot{\beta}}$, would in general be different in such a case, corresponding to geometrically different paths in coordinate space.

The functions $g^{i\alpha\dot{\beta}}$ are determined by the parameters x^i; $i = 1, 2, 3, 4$. Defining

$$2g^{ik} = g^{i\alpha\dot{\beta}} g^k{}_{\alpha\dot{\beta}} \tag{151}$$

gives a Riemannian geometry with g^{ik} as metric tensor. The geometry is determined by the six functions α, β, γ, δ, λ, μ, of the four parameters x^i. Evidently, then, we have attained our objective, which was to specify the geometry in terms of six functions of the coordinates.

To obtain a full quantization of both particles and geometry, we consider all paths in coordinate space for the particles (i.e., all spinor chains) and all functions α, β, γ, δ, λ, μ, for the geometry. There are no subsidiary conditions in such a formulation.

It remains to consider how to interpret a transformation of coordinates $x'^i = x'^i(x)$. In place of (148) we must use

$$x'^i{}_{(n)} = \sum_{r=1}^{n} \frac{\partial x'^i}{\partial x^k} g_{(r)}{}^{k\alpha\dot{\beta}} a^{(r)}{}_{\alpha\dot{\beta}}, \tag{152}$$

and also invert the coordinate relationship to

$$x^i = x^i(x'). \tag{153}$$

Although more roundabout, the procedure is essentially the same as before. Suppose x'^{i}_{n} is known. From (153) obtain $x^{i}_{(n)}$, and use (149) to determine $g_{(n+1)}{}^{i\alpha\dot{\beta}}$. Then make the next step in the spinor chain to determine $x'^{i}_{(n+1)}$ from (152). The spinor chain itself is not affected by the change of coordinates.

To interpret the above procedure, we note that, since α, β, γ, δ, λ, μ, are given in terms of x^{i}, we have to return from x'^{i} to x^{i} through (153) in order to determine the matrix U. The $g^{i\alpha\dot{\beta}}$ resulting from (149) must then be subject to the usual vector transformation before being used in (153).

9

FURTHER CONSIDERATIONS OF COSMOLOGY

§9.1. DIFFICULTIES IN THE PREVIOUS TREATMENT (LECTURE 23)

The discussion of Chapter 6 contained three unsatisfactory features:

(1) The Friedmann cosmologies are inconsistent with the absorber theory of electrodynamics. We shall not discuss this issue here, but we will return to it at the end of this chapter.

(2) When considered in the Minkowski conformal frame, the Friedmann cosmologies are seen to result from the time-derivatives of the mass field. The discussion at the end of Chapter 7 showed that these time-derivatives arise from the broken ends of the paths of particles which occur at the origin of the universe, a situation no different in principle from that in the steady-state model. It hardly makes logical sense to object to the latter model because it requires creation of matter (i.e., broken ends) and at the same time to accept the broken ends of the Friedmann models.

(3) The form of the gravitational equations obtained in Chapter 6 follows along lines similar to the derivation of the canonical tensor in the electromagnetic theory. Thus in (104) of Chapter 6 the equations are symmetric in the advanced and retarded masses. By using the mass-response condition

(115) of that chapter, we were able to convert (104) to the form (117) involving only retarded masses. Then we were able to choose a particular conformal frame in which the retarded mass was constant, and in this frame (117) reduced to the Einstein equations.

We come now to an important point, glossed over in the previous discussion of the Friedmann cosmologies, namely, that these models do not satisfy the response condition (115) of Chapter 6. It follows that, in addition to the unsatisfactory points mentioned above, there is an actual inconsistency in the treatment of these models. To obtain the lefthand side of (115) for the Minkowski frame, we use

$$G^{(\mathrm{adv})}(X, B) = \frac{1}{4\pi}\,\delta(\tau_X - \tau_B + |\boldsymbol{x}_X - \boldsymbol{x}_B|). \tag{1}$$

Working in terms of the Einstein–de Sitter model for simplicity, with a uniform density L^{-3} of particles, we can fairly easily see that

$$M^{(\mathrm{adv})}{}_{\tau > T}(X) = \frac{\lambda L^{-3}}{4\pi} \int_0^{\pi} \sin\theta\, d\theta \int_0^{2\pi} d\varphi \int_{T-\tau_X}^{\infty} r\, dr, \tag{2}$$

which is divergent. On the other hand, the terms on the righthand side of (115) are easily seen to be convergent.

§9.2. THE OTHER HALF OF THE UNIVERSE

We shall now show that both the second and the third of the above points can be resolved by dispensing with an origin of the universe. For Minkowski space, there is no reason why the universe should not exist in the lower half of the plane as well as in the upper half. The conformal transformation function Ω from Minkowski space

$$ds^2 = d\tau^2 - dr^2 - r^2(d\theta^2 + \sin^2\theta\, d\varphi^2) \tag{3}$$

to Robertson-Walker space

$$ds^{*2} = dt^2 - Q^2(t)\left[\frac{dr^2}{1 - kr^2} + r^2(d\theta^2 + \sin^2\theta\, d\varphi^2)\right], \qquad k = 0, \pm 1, \tag{4}$$

is singular at $\tau = 0$. This prevents $\tau < 0$ in Minkowski space from having an analogue in Robertson-Walker space, the implication being that, by our working in Robertson-Walker space, half the physical universe becomes suppressed.

This situation recalls the old horizon difficulty of de Sitter's first cosmological model. Instead of working with $k = 0$, $Q = \exp(Ht)$, in (4), de Sitter obtained his model in the form

$$ds^{*2} = (1 - H^2R^2)\,dT^2 - \frac{dR^2}{1 - H^2R^2} - R^2(d\theta^2 + \sin^2\theta\, d\varphi^2), \qquad (5)$$

where $H^2 = \frac{1}{3}\Lambda$, Λ being the cosmical constant. The *coordinate* transformation relating (5) to the Robertson-Walker form of the line element,

$$ds^{*2} = dt^2 - \exp(2Ht)[dr^2 + r^2(d\theta^2 + \sin^2\theta\, d\varphi^2)], \qquad (6)$$

is

$$t = T + \frac{1}{2H}\ln(1 - H^2R^2), \qquad R = r\exp(Ht). \qquad (7)$$

The form (5) is limited in the R, T, plane to the strip $R < H^{-1}$, which maps into the region of the r, t, plane confined between the curve

$$r = H^{-1}\exp(-Ht) \qquad (8)$$

and the t-axis. Yet we do not restrict the r, t, coordinates to such a region when the Robertson-Walker line element (6) is used. The r, t, coordinates are then permitted to take any values in the half plane $-\infty < t < \infty, 0 \leq r$. We pass to the R, T, coordinates back through the transformation (7), noticing that the transformation is singular on the curve (8), and taking the view that the restriction on R is only a mathematical construct. Such coordinate singularities are well-known in general relativity, and have no physical significance.

In a similar way we now take the view that the singularity at $t = 0$ in the Robertson-Walker space is only a mathematical construct, due to an unfortunate choice of the conformal frame. There is a second half to the universe, which appears without difficulty when the Minkowski frame is used.

The hyperplane $\tau = 0$ must still have exceptional significance. To see what this might be, consider the interaction

$$\int\int \tilde{G}(A, B)\, da\, db \qquad (9)$$

between particles a and b. This interaction formed the basis of the gravitational theory of Chapter VI. It is irrelevant whether we consider the propagator $\widetilde{G}(A, B)$ to couple positively or negatively to da at point A, so long as we adopt the same sign for the coupling to db at point B. The sign of the coupling can be regarded as being determined by the sign of the constant λ in the action (70) of Chapter VI. Since the action depends quadratically on λ, a switch of sign evidently has no effect on the action itself. The signs of the mass fields (80) and (81) of Chapter VI are changed, but since the mass fields appear quadratically in the gravitational equations this change also has no effect. And the $T_{(\mathrm{mat})}{}^{ik}$ term (85) of Chapter VI is unchanged, since $m(A)$ in this expression again involves λ twice, once in the mass field and once in the definition of $m(A)$ in terms of the mass field, as in (88) of Chapter VI.

The situation would be different if the sign of λ were switched at A but not at B. The sign of the action would then be changed, and this change would affect the behavior of particle a in an electromagnetic field. However, such a procedure would be unsymmetric with respect to particles a and b. It is possible to maintain symmetry, however, and yet to make use of the idea of sign change in the mass interaction, by introducing the following rule:

The sign of the coupling is the same as the sign of τ.

For $\tau_A > 0$, $\tau_B > 0$, we have the situation we have considered hitherto. For $\tau_A < 0$, $\tau_B < 0$, we have the equivalent situation with negative coupling at both A and B. For $\tau_A > 0$, $\tau_B < 0$, however, we have positive coupling at A and negative coupling at B—and *vice versa* if $\tau_A < 0$, $\tau_B > 0$. This new rule is illustrated in Figure 9.1.

On this basis, consider the mass field of particle a,

$$m^{(a)}(X) = \int \widetilde{G}(X, A)\lambda(A)\, da, \tag{10}$$

in which $\lambda(A) = \pm\epsilon$ according as $\tau_A \gtrless 0$, ϵ being a positive constant. Working in the Minkowski frame for the Einstein–de Sitter model, let A_0 be the point at $\tau = 0$ on the path of particle a. Draw the light cone of A_0 as in Figure 9.2. In the forward half of this cone, the mass field (10) is $\epsilon/4\pi r$, where r is the three-dimensional spatial distance of point X from the particle. In the backward half of the cone, the field is $-\epsilon/4\pi r$, and outside the cone the field is zero. For $\tau_X > 0$, it follows that the only particles contributing to the total mass field $M(X) = \Sigma_a\, m^{(a)}(X)$ are those with points A_0 lying within the backward half of the light cone at X. For $\tau_X < 0$, $M(X)$ is similarly

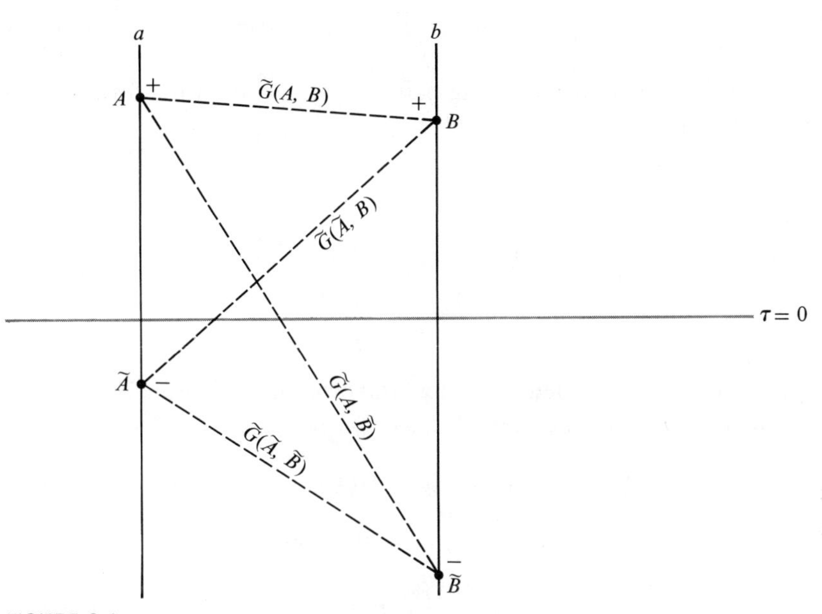

FIGURE 9.1
There are four propagators, $\tilde{G}(A,B)$, $\tilde{G}(A,\tilde{B})$, $\tilde{G}(\tilde{A},B)$, $\tilde{G}(\tilde{A},\tilde{B})$, relating the points A, $\tilde{A}$, on the path of particle a to the points B, $\tilde{B}$, on the path of particle b. With $t_A > 0$, $t_B > 0$, $t_A < 0$, $t_B < 0$, the propagators couple to the particles with positive sign at A, B, and with negative sign at $\tilde{A}$, $\tilde{B}$.

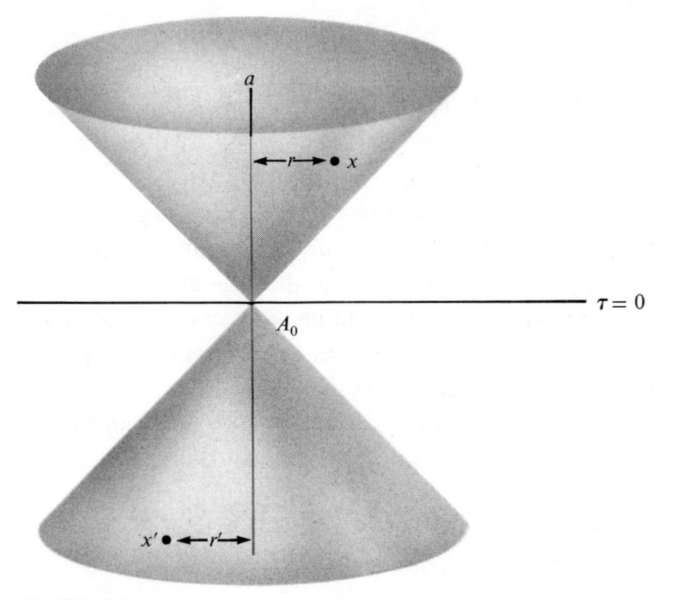

FIGURE 9.2
The mass field due to particle a at point X in the forward half of the cone from A_0 is $e/(4\pi r)$. At point X' in the backward half of the cone, the mass field is $-e/(4\pi r')$.

determined by the particles having points A_0 lying within the forward half of the light cone at X.

The above rule therefore leads to

$$M(X) = \begin{cases} \frac{1}{2}\epsilon L^{-3}\tau_X{}^2, & \tau_X > 0, \\ -\frac{1}{2}\epsilon L^{-3}\tau_X{}^2, & \tau_X < 0, \end{cases} \tag{11}$$

where L^{-3} is the particle density in the Minkowski frame for the Einstein–de Sitter model. Thus for the inertial mass of a particle we get

$$m(X) = \lambda(X)M(X), \tag{12}$$

and provided we keep to our rule for λ, the sign of m is always positive,

$$m(X) = \frac{1}{2}\epsilon^2 L^{-3}\tau_X{}^2, \tag{13}$$

whatever the sign of τ_X. Except for the trivial change of writing ϵ rather than λ^2 for the coupling constant in (13), this is the same as the inertial mass used in Chapter 7.

§9.3. PARTICLES OF NEGATIVE MASS

Why can we not have particles with the sign of λ inverted, $\lambda = -\epsilon$ for $\tau > 0$, or $\lambda = +\epsilon$ for $\tau < 0$? Such particles would have negative mass $-m(X)$, $m(X)$ still being given by (13). It hardly seems satisfactory to exclude this possibility by a purely axiomatic appeal to the above rule, since otherwise the geometrical surface $\tau = 0$ would be represented as the physical cause of the switch in sign of λ. Rather must we argue the opposite way—that the switch in λ determines the surface $\tau = 0$.

Two particles, a of positive mass and b of negative mass, have an energy-momentum tensor

$$\int \frac{\delta_4(X, A)}{\sqrt{-g(A)}} m(A) \frac{da^i}{da}\frac{da^k}{da}\, da - \int \frac{\delta_4(X, B)}{\sqrt{-g(B)}} m(B) \frac{db^i}{db}\frac{db^k}{db}\, db. \tag{14}$$

Such a pair would be capable of annihilation without release of energy. Particles of negative mass are therefore not antiparticles, since particles and antiparticles always annihilate with nonzero energy release. A general assembly of particles, some of positive mass, others of negative mass, would tend to evolve until only particles of one kind remained, namely, the kind that happened to be in the majority in the first place. Here we have a reason why the sign of the coupling constant may be expected to be the same throughout substantial regions of the universe. Different regions could have different signs, regions with $\lambda = +\epsilon$ giving rise to positive contributions to the total mass field $M(X)$, and regions with $\lambda = -\epsilon$ giving rise to negative contributions. The case considered above, $\lambda = +\epsilon$ for $\tau > 0$, $\lambda = -\epsilon$ for $\tau < 0$, is a simple example of this, with the universe separated into only two regions. In principle, however, the universe may be very much more complex than such a simple model.

The general implications of a multiregion universe will be considered in §9.5. For now, we return to the question formulated above. Taking our region of the universe to have $\lambda = +\epsilon$, what would be the properties of a particle with $\lambda = -\epsilon$, assuming such a particle to exist temporarily in our region? Instead of being propagated quantum-mechanically by K_+, i.e., with positive-energy states going forward in time and negative-energy states going backward, such a particle would reveal the opposite situation, with negative-energy states going forward and positive-energy states going backward. The appropriate propagator would be

$$K_-(2;1) = \begin{cases} \sum\limits_{E_n < 0} u_n(2)\bar{u}_n(1), & t_2 > t_1, \\ \\ -\sum\limits_{E_n > 0} u_n(2)\bar{u}_n(1), & t_2 < t_1. \end{cases} \tag{15}$$

In terms of Hankel functions,

$$K_+(2;1) = (\nabla\!\!\!\!/_2 - im)\left[\frac{1}{4\pi}\,\delta(s^2{}_{21}) - \frac{m}{8\pi s_{21}}\,H_1^{(2)}(ms_{21})\right], \tag{16}$$

$$K_-(2;1) = -(\nabla\!\!\!\!/_2 - im)\left[\frac{1}{4\pi}\,\delta(s^2{}_{21}) - \frac{m}{8\pi s_{21}}\,H_1^{(1)}(ms_{21})\right]. \tag{17}$$

The additional minus sign in K_- is related to the time sense. Inversion, $\tau \to -\tau$, transfers the minus sign to K_+. Otherwise the difference in behavior of a negative-mass particle from one of positive mass is determined only by the nature of the Hankel functions. In (124) of Chapter 8, we had

$$H_1^{(2)}(ms_{21}) \simeq \left(\frac{2}{\pi m s_{21}}\right)^{1/2} \exp\left[-i\left(ms_{21} - \frac{3\pi}{2}\right)\right], \qquad ms_{21} \gg 1. \quad (18)$$

The corresponding asymptotic expression for $H_1^{(1)}(ms_{21})$ is

$$H_1^{(1)}(ms_{21}) \simeq \left(\frac{2}{\pi m s_{21}}\right)^{1/2} \exp\left[i\left(ms_{21} - \frac{3\pi}{2}\right)\right], \qquad ms_{21} \gg 1. \quad (19)$$

The quantity ms_{21} in these expressions arises in the free-particle case from evaluating $\int mds$. The phase

$$-ms_{21} = -\int mds$$

determines the classical action for a positive-mass particle, as we saw in §8.5. The corresponding phase in (19)

$$+\int mds, \quad (20)$$

also determines the classical action. The sign change here characterizes a particle of negative mass. The contribution to the gravitational field is opposite to that of a positive-mass particle. There is also a switch of sign in the charge-to-mass ratio, leading to a different orbit in a prescribed electromagnetic field. On the other hand, the orbit in a prescribed gravitational field is the same as for a positive-mass particle.

The appearance of the K_- propagator is interesting. It fills a logical gap in the usual theory. Particles are propagated by K_+ in our region of the universe, but there can be other circumstances in which K_- is the appropriate propagator. By these considerations the universe takes a more symmetric form.

§9.4. OSCILLATING UNIVERSE AND BLACK HOLES

The Friedmann cosmological model with $k = +1$ when considered in the Robertson-Walker conformal frame does not expand indefinitely. It expands to a minimum density and then falls back, ultimately reaching a metric singularity. As we saw in Chapter 2, the interior solution for an infalling spherically symmetric uniform local body has the same property, and indeed has an identical mathematical structure. No recourse within the usual gravitational theory has been found for preventing these singularities. Work by Penrose (1965) and by Hawking and Penrose (1969) shows that it is impossible to prevent singularities from arising in these problems—a result which is of serious concern to most cosmologists. This situation is easily resolved, however, in terms of the ideas we are discussing here.

The Friedmann model with $k = +1$ was discussed in Chapter 7 in relation to the Minkowski conformal frame. The expanding phase of the model was represented by the expansion to infinity of a finite distribution of material. Some subtlety was involved in understanding the connection of the expanding and contracting phases—a connection more easily discussed with respect to the Robertson-Walker frame. The Minkowski frame is much better suited, however, to the contracting phase, since in the Minkowski frame *there is no metric singularity*. The finite cloud of Chapter 7 contracts back* to its initial configuration at $\tau = 0$, whereupon it expands again, now with decreasing τ, to $\tau \to -\infty$. This second expansion is similar to the first expansion from $\tau = 0$ to $\tau \to +\infty$. As in the previous expansion phase, the cloud turns around at $\tau \to -\infty$, contracting back to $\tau = 0$, and beginning a new cycle of expansion from $\tau = 0$ to $\tau \to +\infty$. Evidently, by extending the Minkowski frame to include the lower half of the plane, we have arrived at an oscillatory model. This goal has often been sought, but has never been attained within the scope of the usual theory. It is not hard to understand the reason for this failure.

So long as the mass field $M(X)$ is nonzero, it is possible to choose a conformal frame in which M is constant. With respect to this frame, particle masses are constant and the conformally invariant gravitational theory then reduces to the Einstein theory. But with some regions of the universe making positive contributions to $M(X)$ and other regions making negative contributions, there can be surfaces of zero M. An attempt to choose a frame with M constant then leads to artificial metric singularities at these surfaces. It is clear that the usual theory, starting with the assumption of constant particle masses, must break down at all such surfaces, and will therefore fail to make proper connections across them. This failure is why a satisfactory development of the $k = +1$ Friedmann model in terms of repeated oscillations has never been found within the scope of the usual theory, and why an oscillatory interpretation appears without difficulty in a conformally invariant theory.

The same considerations apply to the implosion of a local body—the so-called black-hole problem. The metric singularity of the usual theory is

---

*In the contracting phase we adopt the convention of measuring the time τ backward.

an artifact of the assumption of constant particle masses. Just as in the $k = +1$ cosmological model, the implosion phase is followed by expansion—the black hole becomes a white hole. However, the white hole cannot appear with respect to the same external observer who witnessed the infall of the body into a black hole. It is an interesting problem, not investigated here, to determine the relation of external observers, one witnessing the black hole development and the other witnessing the white hole. It is an attractive conjecture, requiring proof, that two such observers must lie on opposite sides of a surface of zero mass.

§9.5. NEW COSMOLOGICAL MODELS (LECTURE 24)

Like the situation in electrodynamics, where we regard the universe as being made up of equal numbers of positive and negative charges, we can think in terms of equal numbers of $\pm$ couplings of particles to the mass field $M(X)$. In electrodynamics we cannot consider an arbitrarily assigned four-dimensional distribution of the charges. We can certainly assign a distribution on a three-dimensional spacelike surface, but the charges then give rise to an electromagnetic field that imposes conditions on the trajectories of the charges—in other words, only certain four-dimensional distributions of the charges are consistent with the electrodynamic equations.

Similarly, we cannot consider an arbitrarily assigned four-dimensional distribution of $\pm$ mass couplings, because an arbitrary distribution would not in general be consistent with the gravitational equations. Just as the electrodynamic equations prevent large-scale charge separations from generating exceedingly large potentials, so we may expect the gravitational equations to prevent exceedingly large mass values from being developed. However, because of the comparative weakness of gravitation, it is not necessary for positive contributions to the mass field to be as closely compensated by negative contributions as $+$ and $-$ charges are compensated in electrodynamics. We can think in terms of positive aggregations, and of negative aggregations, on a scale that we usually think of as "cosmological"—that is, on at least the scale of H^{-1}. Although the aggregations are "local" in relation to the whole universe, they are nevertheless very large compared to the separation of galaxies, for example.

With positive and negative aggregations distributed to give equal average densities of particles with $+$ and $-$ couplings, the value of $M(X)$ is as likely to be negative as positive, which implies the existence of at least one surface with $M = 0$. Such a surface is conventionally associated in cosmology with the origin of the universe. However, the theory here does not require such an origin, nor does it require the particular zero surface indicated by astronomical observations to be unique. There could be many such surfaces.

Consider a region of spacetime close to a surface $M = 0$, taking either the surface, or the relevant portion of it, to be spacelike. Setting up comoving coordinates in the usual way, the line element can be written in the form

$$ds^2 = dt^2 + g_{\mu\nu}\, dx^\mu\, dx^\nu; \qquad \mu, \nu = 1, 2, 3. \tag{21}$$

Provided the region is not too large, the sections $t =$ constant can be taken to be isotropic and homogeneous, and (21) can be expressed in the Robertson-Walker form, whence by Chapter 7 a conformal transformation can be found to reduce the line element to the Minkowski form,

$$ds^{*2} = d\tau^2 - dr^2 - r^2(d\theta^2 + \sin^2\theta\, d\varphi^2). \tag{22}$$

Expressing $M(\tau)$ locally as a power-series expansion in τ, we have

$$M(\tau) = \frac{1}{2} B\tau + \frac{1}{2}\lambda L^{-3}\tau^2 + \dots, \tag{23}$$

where L^{-3} is the particle density and B is a constant with the dimensionality L^{-2}. The form of the τ^2 term in (23) follows from the wave equation for $M(X)$,

$$\Box_X M + \frac{1}{6} RM = \lambda \sum_a \int \frac{\delta_4(X, A)}{\sqrt{-g(A)}}\, da, \tag{24}$$

all particles in the region in question being taken to have the same λ.

Next we ask for the restrictions imposed by the gravitational equations. In the general form given in Chapter 6, these are

$$K\left(R_{ik} - \frac{1}{2} g_{ik} R\right) = -3(T_{ik} + \Phi_{ik}) + g_{ik}\Box K - K_{;ik}, \tag{25}$$

$$K = \frac{1}{2} M^{(\mathrm{ret})} M^{(\mathrm{adv})}, \qquad M = \frac{1}{2}[M^{(\mathrm{ret})} + M^{(\mathrm{adv})}], \tag{26}$$

$$\Phi_{ik} = -\frac{1}{2}[M_i^{(\mathrm{ret})} M_k^{(\mathrm{adv})} + M_k^{(\mathrm{ret})} M_i^{(\mathrm{adv})} - g_{ik} g^{pq} M_p^{(\mathrm{ret})} M_q^{(\mathrm{adv})}], \tag{27}$$

$$T_{ik}(X) = \sum_a \int \frac{\delta_4(X, A)}{\sqrt{-g(A)}} \lambda(A) M(A) \frac{da_i}{da} \frac{da_k}{da}\, da. \tag{28}$$

Because we wish to work here in general terms, we prefer to start with (104) of Chapter 6, involving $M^{(\mathrm{ret})}$, $M^{(\mathrm{adv})}$, rather than with the Einstein equations, which were derived from (104) with the aid of the response condition (115) of Chapter 6. Since $M^{(\mathrm{ret})}$, $M^{(\mathrm{adv})}$, are both solutions of (24), we can write

$$M^{(\mathrm{ret})} = a + b\tau + \frac{1}{2}\lambda L^{-3}\tau^2 + \dots, \tag{29}$$

$$M^{(\mathrm{adv})} = -a + (B - b)\tau + \frac{1}{2}\lambda L^{-3}\tau^2 + \dots, \tag{30}$$

where a, b, are constants with dimensionalities L^{-1}, L^{-2}, respectively. The lefthand side of (25) is zero in the Minkowski frame. Moreover, only the 44 component of T_{ik} is nonzero, since the coordinates are comoving. It follows that the coefficient of g_{ik} on the righthand side of (25) must be zero,

$$\Box K - \frac{3}{2} g^{pq} M_p{}^{(\mathrm{ret})} M_q{}^{(\mathrm{adv})} = 0. \tag{31}$$

Working to the second order in τ^2, we have

$$\begin{aligned} \Box K - \frac{3}{2} g^{pq} M_p{}^{(\mathrm{ret})} M_q{}^{(\mathrm{adv})} &= \frac{1}{2}\{M^{(\mathrm{ret})}\Box M^{(\mathrm{adv})} + M^{(\mathrm{adv})}\Box M^{(\mathrm{ret})} \\ &\quad - \dot{M}^{(\mathrm{ret})}\dot{M}^{(\mathrm{adv})}\} \\ &= -\frac{1}{2} b(B - b), \end{aligned} \tag{32}$$

the terms in τ and τ^2 vanishing identically. Hence the condition (31) requires the coefficient of τ in $M^{(\mathrm{ret})}$ or in $M^{(\mathrm{adv})}$ to be zero. Either we have

$$\begin{aligned} M^{(\mathrm{ret})} &= a + B\tau + \frac{1}{2}\lambda L^{-3}\tau^2 + \dots, \\ M^{(\mathrm{adv})} &= -a + \frac{1}{2}\lambda L^{-3}\tau^2 + \dots, \end{aligned} \tag{33}$$

or

$$\begin{aligned} M^{(\mathrm{ret})} &= a + \frac{1}{2}\lambda L^{-3}\tau^2 + \dots, \\ M^{(\mathrm{adv})} &= -a + B\tau + \frac{1}{2}\lambda L^{-3}\tau^2 + \dots. \end{aligned} \tag{34}$$

The gravitational equations applied to the homogeneous cosmological situation carry us no further than this, since the contracted form of (25) leads to (24), which has been used already to arrive at the terms in τ^2. To proceed, we must either use the response condition (115) of Chapter 6, in which case

$$M^{(\mathrm{ret})} = M^{(\mathrm{adv})}, \tag{35}$$

and

$$a = 0, \qquad B = 0, \tag{36}$$

or we must accept the possibility that a, B, could be nonzero. In the first case, we have the Einstein–de Sitter model once again, but now applying only sufficiently close to the surface $M = 0$—there is no requirement here for $M^{(\mathrm{ret})}$ and $M^{(\mathrm{adv})}$ to be equal to $\frac{1}{2}\lambda L^{-3}\tau^2$ throughout the whole universe. In the second case, we have possibilities for new cosmological models, which we shall consider in the rest of this lecture.

Neglecting terms of higher order than τ^2 in the expressions for $M^{(\mathrm{ret})}$, $M^{(\mathrm{adv})}$, and defining G in the usual way, we have

$$8\pi G = \frac{3}{K} = \frac{6}{M^{(\mathrm{ret})}M^{(\mathrm{adv})}}. \tag{37}$$

Since (37) is symmetric in $M^{(\mathrm{ret})}$, $M^{(\mathrm{adv})}$, the forms given in (33) do not contain possibilities different from those contained in (34). Using (34), and defining

$$\mu = BL^2, \qquad \nu = aL, \tag{38}$$

we obtain

$$8\pi G = 6L^2\left(-\nu^2 + \mu\nu L^{-1}\tau + \frac{1}{2}\lambda\mu L^{-3}\tau^3 + \frac{1}{4}\lambda^2 L^{-4}\tau^4\right)^{-1}. \tag{39}$$

The righthand side here can be positive or negative according to the values of λ, μ, ν, τ. When G and the inertial mass m, given by

$$m = \lambda M = \frac{1}{2}\lambda[M^{(\mathrm{ret})} + M^{(\mathrm{adv})}] = \frac{1}{2}L^{-1}(\lambda\mu L^{-1}\tau + \lambda^2 L^{-2}\tau^2), \tag{40}$$

have the same sign, the particles attract each other, but when G and m have opposite signs, the particles repel. Again depending on the values of λ, μ, ν, τ, we can have either case.

Switching the signs of μ and τ leaves (39) and (40) unchanged. Hence no possibilities arise for $\tau < 0$ that are not present for $\tau > 0$. It is sufficient therefore to confine attention to $\tau > 0$. The model is determined by the numerical values of the dimensionless parameters λ, μ, ν. We shall restrict the following discussion to examples of special interest.

A. $\nu = 0$, $\lambda\mu > 0$

G and m are both positive; so the particles attract each other in this case. The character of the model is perhaps best understood by transforming conformally to a frame in which the inertial mass becomes independent of τ. Taking

$$\Omega = \frac{\lambda L^{-1}}{2m_0}(\mu L^{-1}\tau + \lambda L^{-2}\tau^2), \qquad \tau > 0, \tag{41}$$

we get

$$m^* = \Omega^{-1} m = m_0, \tag{42}$$

$$G^* = \Omega^2 G = \frac{3}{4\pi}\frac{\lambda^2}{m_0{}^2}\frac{(\mu/\lambda + \tau/L)^2}{2\mu\tau/\lambda L + \tau^2/L^2}. \tag{43}$$

When

$$\tau/L \gg \mu/\lambda, \tag{44}$$

we have

$$\Omega \simeq \frac{\lambda^2}{2m_0} L^{-3}\tau^2, \tag{45}$$

$$G^* = \frac{3\lambda^2}{4\pi m_0{}^2}. \tag{46}$$

The line element is transformed by (45) from the Minkowski frame to

$$\begin{aligned} ds^{*2} &= \frac{\lambda^4}{4m_0{}^2} L^{-6}\tau^4[d\tau^2 - dr^2 - r^2(d\theta^2 + \sin^2\theta\, d\varphi^2)] \\ &= dt^2 - Q^2(t)[dr^2 + r^2(d\theta^2 + \sin^2\theta\, d\varphi^2)], \end{aligned} \tag{47}$$

where

$$Q(t) = \left(\frac{2\lambda^2 L^{-3}}{m_0}\right)^{1/3}\left(\frac{3t}{2}\right)^{2/3}. \tag{48}$$

The forms (46), (47), (48), are the same as those obtained for the Einstein–de Sitter model in Chapter 7, except for an interchange of starred and unstarred quantities due to our working here from the Minkowski frame instead of from the Robertson-Walker frame.

The situation is very different from the Einstein–de Sitter model, however, when

$$\frac{\tau}{L} \ll \frac{\mu}{\lambda}. \tag{49}$$

In place of the relation (45) to (48), we then have

$$\Omega \simeq \frac{\lambda\mu}{2m_0} L^{-2}\tau, \tag{50}$$

$$G^* \simeq \frac{3}{8\pi}\frac{\lambda\mu}{m_0^2}\cdot\frac{L}{\tau}, \tag{51}$$

$$\begin{aligned} ds^{*2} &= \frac{\lambda^2\mu^2}{4m_0^2} L^{-4}\tau^2[d\tau^2 - dr^2 - r^2(d\theta^2 + \sin^2\theta\, d\varphi^2)], \\ &= dt^2 - Q^2(t)[dr^2 + r^2(d\theta^2 + \sin^2\theta\, d\varphi^2)], \end{aligned} \tag{52}$$

where

$$t = \frac{\lambda\mu}{4m_0} L^{-2}\tau^2, \tag{53}$$

$$Q(t) = \left(\frac{\lambda\mu t}{m_0 L^2}\right)^{1/2}. \tag{54}$$

Instead of G^* being independent of time, as in (46), we have

$$G^* = \frac{3}{16\pi}\left(\frac{\lambda\mu}{m_0}\right)^{3/2}\frac{1}{m_0 t^{1/2}}. \tag{55}$$

The definitions of the instantaneous cosmological parameters H, q_0,

$$H = \frac{\dot{Q}}{Q}, \qquad q_0 = -\frac{1}{H^2}\frac{\ddot{Q}}{Q}, \tag{56}$$

give

$$t = \frac{1}{2}H^{-1}, \qquad q_0 = +1, \tag{57}$$

compared to

$$t = \frac{2}{3} H^{-1}, \qquad q_0 = +\frac{1}{2}, \tag{58}$$

for the Einstein–de Sitter model.

The mass density in the Robertson-Walker frame is given by

$$\sigma = m_0 L^{-3} Q^{-3}. \tag{59}$$

Using (46), (48), (58), for $\tau/L \gg \mu/\lambda$, we get

$$\sigma = \frac{3H^2}{8\pi G^*}, \tag{60}$$

whereas using (54), (55), (57), for $\tau/L \ll \mu/\lambda$ gives

$$\sigma = \frac{3H^2}{4\pi G^*}. \tag{61}$$

Summarizing, near enough to the surface of zero mass we have a model with variable gravity, of the kind first proposed by Dirac (1937, 1938b). Such models have interesting astrophysical and geophysical properties. They have been considered by Jordan (1949), by Dicke (1964, p. 142), and recently by Hoyle and Narlikar (1972). However, as τ increases, there is an eventual transition to the usual Einstein–de Sitter model.

B. $\nu \neq 0, \quad \lambda\mu > 0$

Start with the previous case, and let ν increase slightly from zero. The only change will be for τ very close to zero, where

$$G \simeq -\frac{3}{4\pi}\frac{L^2}{\nu^2}. \tag{62}$$

With G negative and with the inertial mass (40) remaining positive, the particles repel each other. Indeed, since the denominator of (39) must have a zero at a finite positive value of τ, very large repulsive forces arise close to the surface of zero mass. Such forces evaporate any small-scale condensations that might be present at $\tau = 0$, and hence serve to produce a homogeneous situation.

There are observational indications, particularly from the well-known microwave background radiation, that the universe must have been exceedingly homogeneous close to the surface of zero mass. It is not easy to understand within the compass of the usual theory why this should be so. Hence this case of small ν has considerable interest in providing a possible resolution of this problem.

More complex properties arise for larger values of ν, but these will not be considered here. Enough has been said to show that a wide range of new cosmological situations, not found in the usual theory, can arise if we are permitted to make appropriate choices for the dimensionless parameters λ, μ, ν. This possibility for obtaining new models turns on whether the response condition (115) of Chapter 6 must be applied or not. This condition leads to $M^{(\mathrm{ret})} = M^{(\mathrm{adv})}$ and thence to $\mu = 0$, $\nu = 0$, thereby removing the new class of models. From an astrophysical point of view, it is desirable to find new cosmological possibilities. On the other hand, the response condition (115) of Chapter 7, with its analogy to the electromagnetic response condition, has deeper theoretical interest. How are we to decide between these alternatives? In §9.6 we shall consider the observational implication of changing gravity, making explicit reference to the case discussed above. The issue to be decided is whether in our particular region of spacetime the cosmological line element could be determined by (52), (54), with G^* given by (55). We have to ask whether this possibility is consistent with all available data.

§9.6. IMPORTANCE OF THE CLASSIC TESTS OF GENERAL RELATIVITY (LECTURE 25)

We turn immediately to the issue just raised, our aim being to discuss here the possibility that the locality observed by the astronomer corresponds to the case $\nu = 0$, $\lambda\mu > 0$, $\tau \ll \mu L/\lambda$. The relevant equations in the Minkowski frame were

$$8\pi G \simeq \frac{12L^5}{\lambda\mu\tau^3}, \tag{63}$$

$$m = \frac{1}{2}\lambda[M^{(\mathrm{ret})} + M^{(\mathrm{adv})}] \simeq \frac{1}{2}\lambda\mu L^{-2}\tau, \tag{64}$$

with either

$$M^{(\mathrm{ret})} \simeq \mu L^{-2}\tau, \qquad M^{(\mathrm{adv})} \simeq \frac{1}{2}\lambda L^{-3}\tau^2, \tag{65}$$

or

$$M^{(\mathrm{ret})} \simeq \frac{1}{2}\lambda L^{-3}\tau^2, \qquad M^{(\mathrm{adv})} \simeq \mu L^{-2}\tau. \tag{66}$$

The conformal transformation

$$\Omega = \frac{\lambda\mu}{2m_0} L^{-2}\tau \tag{67}$$

gave

$$ds^{*2} = dt^2 - \frac{\lambda\mu t}{m_0 L^2}[dr^2 + r^2(d\theta^2 + \sin^2\theta\, d\varphi^2)], \tag{68}$$

where

$$t = \frac{\lambda\mu}{4m_0} L^{-2}\tau^2, \tag{69}$$

together with

$$m^* = m_0, \qquad G^* = \frac{3}{16\pi}\left(\frac{\lambda\mu}{m_0}\right)^{3/2} \frac{1}{m_0 t^{1/2}}. \tag{70}$$

For observations of galaxies, the line element (68) gives $H^{-1} = 2t$, $q_0 = +1$, which agree well with recent work by Sandage (1972a, b). The determination $H^{-1} \simeq 20 \times 10^9$ years in this work gives $t \simeq 10 \times 10^9$ years, which is in satisfactory agreement with independent estimates for the age of our own galaxy. Thus there seems to be no conflict between the above scheme and the results of "observational cosmology."

The variation of G^* with t, according to (70), has many astrophysical and geophysical consequences, which also appear to be in satisfactory qualitative agreement with a wide range of data—so much so that it would not be unreasonable to conclude that the data give quite firm support to (70). Indeed an analysis of the situation is likely to leave one in enthusiastic support of the above model. It is disconcerting therefore to find the theory in conflict with the classic tests of general relativity, with the bending of light by the Sun, and with the rotation of the perihelion of Mercury. We shall be concerned with discussing these local solar-system problems.

The difficulty does not arise directly from the time variability of G^*, which can be taken in these particular problems to remain constant during the time intervals in question—only a few minutes for the light problem and $\sim 10^2$ years for the perihelion motion. The factor $\lambda\mu t/(m_0 L^2)$ in (68) can also be taken to be constant, and $(\lambda\mu t/m_0 L^2)^{1/2}$ can be absorbed as a scale factor into the r coordinate. We regard the Sun as giving rise to a central field essentially static with respect to the large-scale features of the observed system of galaxies. Taking spherical polar coordinates with $r = 0$ at the solar center, we can write

$$ds^{*2} = e^f\,dt^2 - e^g\,dr^2 - r^2(d\theta^2 + \sin^2\theta\,d\varphi^2) \tag{71}$$

for the local metric, where f, g, are functions of r only. We shall denote their derivatives with respect to r by f', g'. By choosing a suitable conformal frame, it is possible, as we shall show first, to express e^f, e^g, exterior to the Sun in the forms

$$e^f = 1 - \frac{S}{r} + O\left(\frac{1}{r^3}\right), \qquad e^g = 1 + \frac{S}{r} + O\left(\frac{1}{r^2}\right), \tag{72}$$

where S is a constant with the dimensionality of length. This result would seem at first sight to lead in the post-Newtonian approximation to the same behavior as the Einstein theory—and hence to a satisfactory resolution of the present issue. We shall see why this is not so shortly.

Take the conformal frame to be such that $K = \frac{1}{2}M^{(\text{ret})}M^{(\text{adv})}$ is independent of r. Expanding $M^{(\text{ret})}$, $M^{(\text{adv})}$, in the exterior in powers of $1/r$, we then have

$$M^{(\text{ret})} = M_0^{(\text{ret})}\left(1 + \frac{A}{r} + \ldots\right), \qquad M^{(\text{adv})} = M_0^{(\text{adv})}\left(1 - \frac{A}{r} + \ldots\right), \tag{73}$$

where A is a constant with the dimensionality of length and $M_0^{(\text{ret})}$, $M_0^{(\text{adv})}$ are the present-day cosmological values at large r. If (65) holds,

$$\frac{M_0^{(\text{ret})}}{M_0^{(\text{adv})}} = \frac{2\mu}{\lambda} L\tau^{-1} = \left(\frac{\mu^3}{\lambda m_0}\right)^{1/2} t^{-1/2}, \tag{74}$$

whereas $M_0^{(\text{adv})}/M_0^{(\text{ret})}$ is equal to the righthand side of (74) if (66) holds. *It is the circumstance that the righthand side of (74) is not unity in the present model that leads to difficulty with the classic tests of relativity.* When, on the other hand, we use the response condition (115) of Chapter 6, we get $M_0^{(\text{ret})} = M_0^{(\text{adv})}$ and the difficulty is resolved.

With K constant, the gravitational equations exterior to the Sun take the form

$$K\left(R_{ik} - \frac{1}{2}g_{ik}R\right) = -3\Phi_{ik} = \frac{3}{2}[M_i^{(\text{ret})}M_k^{(\text{adv})} + M_k^{(\text{ret})}M_i^{(\text{adv})} - g_{ik}g^{pq}M_p^{(\text{ret})}M_q^{(\text{adv})}], \tag{75}$$

terms of cosmological order being neglected. To the same approximation, we neglect terms on the righthand side of (75) involving time-derivatives. Working to order $1/r^4$, we have

$$-3\Phi_{11} = -\frac{3}{2}A^2 \frac{M_0^{(\mathrm{ret})}}{r^4} M_0^{(\mathrm{adv})}, \tag{76}$$

$$-3\Phi_{44} = -\frac{3}{2}A^2 e^{f-g} \frac{M_0^{(\mathrm{ret})}}{r^4} M_0^{(\mathrm{adv})}. \tag{77}$$

Using known forms for $R_{ik} - \frac{1}{2}g_{ik}R$, after canceling $\frac{1}{2}M_0^{(\mathrm{ret})}M_0^{(\mathrm{adv})}$, we have

$$R_{11} - \frac{1}{2}g_{11}R = -\frac{f'}{r} - \frac{1-e^g}{r^2} = -\frac{3A^2}{r^4}, \tag{78}$$

$$e^{-f+g}\left[R_{44} - \frac{1}{2}g_{44}R\right] = -\frac{g'}{r} + \frac{1-e^g}{r^2} = -\frac{3A^2}{r^4}. \tag{79}$$

From (78), (79),

$$f' + g' = \frac{6A^2}{r^3}, \tag{80}$$

which gives

$$f + g = -\frac{3A^2}{r^2}, \tag{81}$$

since the asymptotic form of (71) at large r is the Minkowski line element. It follows that if we write

$$e^f = 1 - \frac{S}{r} + \frac{B}{r^2} + \dots, \tag{82}$$

then e^g has the form

$$e^g = 1 + \frac{S}{r} + O\left(\frac{1}{r^2}\right). \tag{83}$$

Next we show that $B = 0$. Multiplying (78) by $-re^f$, we can easily see that

$$e^f\left(f' + \frac{1-e^g}{r}\right) = \frac{3A^2}{r^3} + O\left(\frac{1}{r^4}\right). \tag{84}$$

Substituting (82), (83), for e^f, e^g, the lefthand side of (84) is

$$\frac{S}{r^2} - \frac{2B}{r^3} + \frac{1}{r}\left(1 - \frac{S}{r} + \frac{B}{r^2}\right) - \frac{1}{r}\left(1 - \frac{3A^2}{r^2}\right) + O\left(\frac{1}{r^4}\right). \tag{85}$$

Hence (84) gives $B = 0$ and (72) is established.

If planets were to follow geodesic orbits in the conformal frame in question, all would be well: the situation in the post-Newtonian approximation would be the same as in Einstein's theory. However, the equations of motion of a particle, say, a, are

$$\frac{d}{da}\left[m(A)\frac{da^i}{da}\right] + m(A)\Gamma^i_{kl}\frac{da^k}{da}\frac{da^l}{da} = m^i. \tag{86}$$

The orbit is not geodesic because m is not constant.

Since this issue is critically important, it is worth interrupting our discussion to prove (86). The theory being conformally invariant, it is sufficient to obtain (86) in a particular frame, for example, the frame used above with K constant. Since $(R^{ik} - \frac{1}{2}g^{ik}R)_{;k} \equiv 0$, we have

$$\begin{aligned} 2g^{kl}T_{ik;l} &= g^{kl}[M_i^{(\text{ret})}M_k^{(\text{adv})} + M_k^{(\text{ret})}M_i^{(\text{adv})} - g_{ik}g^{pq}M_p^{(\text{ret})}M_q^{(\text{adv})}]_{;l} \\ &= M_i^{(\text{ret})}\Box M^{(\text{adv})} + M_i^{(\text{adv})}\Box M^{(\text{ret})} + g^{kl}M_k^{(\text{ret})}M_{i;\,l}^{(\text{adv})} \\ &\quad + g^{kl}M_k^{(\text{adv})}M_{i;\,l}^{(\text{ret})} - g^{pq}M_{p;\,i}^{(\text{ret})}M_q^{(\text{adv})} - g^{pq}M_p^{(\text{ret})}M_{q;\,i}^{(\text{adv})}. \end{aligned} \tag{87}$$

Because $M^{(\text{ret})}$, $M^{(\text{adv})}$, are scalars it is possible to invert indices, $M_{i;\,l}^{(\text{adv})} = M_{l;\,i}^{(\text{adv})}$ for example, so that the last four terms on the righthand side of (87) cancel in pairs. Hence

$$\begin{aligned} g^{kl}T_{ik;l} &= \frac{1}{2}M_i^{(\text{ret})}\Box M^{(\text{adv})} + \frac{1}{2}M_i^{(\text{adv})}\Box M^{\text{ret}} \\ &= \frac{1}{2}M_i^{(\text{ret})}\left[-\frac{1}{6}RM^{(\text{adv})} + \sum_a \int \frac{\delta_4(X, A)}{\sqrt{-g(A)}}\lambda(A)\,da\right] \\ &\quad + \frac{1}{2}M_i^{(\text{adv})}\left[-\frac{1}{6}RM^{(\text{ret})} + \sum_a \int \frac{\delta_4(X, A)}{\sqrt{-g(A)}}\lambda(A)\,da\right], \end{aligned} \tag{88}$$

where we have used the wave equations satisfied by $M^{(\text{ret})}$, $M^{(\text{adv})}$. Remembering that $K = \frac{1}{2}M^{(\text{ret})}M^{(\text{adv})}$ is being taken constant, and that

$$m = \frac{1}{2}\lambda[M^{(\text{ret})} + M^{(\text{adv})}], \tag{89}$$

it is easy to see that (88) leads to

$$g^{kl}T_{ik;\,l} = m_i \sum_a \int \frac{\delta_4(X, A)}{\sqrt{-g(A)}}\, da. \tag{90}$$

An analysis of $T^{ik}{}_{;k}$, similar to that given at the end of Chapter 6, leads to

$$T^{ik}{}_{;k} = \sum_a \int da \frac{\delta_4(X, A)}{\sqrt{-g(A)}} \left\{ \frac{d}{da}\left[m(A) \frac{da^i}{da} \right] + \Gamma^i{}_{lm} \frac{da^l}{da} \frac{da^m}{da} m(A) \right\}. \tag{91}$$

Thus for (90) to hold at all points X, we require (86) to hold individually for every particle.

The next step is to determine $m(X)$. Writing

$$M^{(\text{ret})} = M_0{}^{(\text{ret})} u(r), \qquad M^{(\text{adv})} = M_0{}^{(\text{adv})} v(r), \tag{92}$$

we have

$$\Box u + \frac{1}{6} Ru = \frac{1}{M_0{}^{(\text{ret})}} \sum_a \int \frac{\delta_4(X, A)}{\sqrt{-g(A)}} \lambda(A)\, da, \tag{93}$$

$$\Box v + \frac{1}{6} Rv = \frac{1}{M_0{}^{(\text{adv})}} \sum_a \int \frac{\delta_4(X, A)}{\sqrt{-g(A)}} \lambda(A)\, da, \tag{94}$$

the summations here being essentially confined to the particles of the Sun. In the exterior we have the expansions

$$u = 1 + \frac{A}{r} + \frac{D}{r^2} + \dots, \qquad v = 1 - \frac{A}{r} + \frac{E}{r^2} + \dots. \tag{95}$$

Working for the moment to order $1/r$, the sources of $\Box v$ must be opposite to those of $\Box u$. By adding (93), (94), and putting $u = 1$, $v = 1$, in the terms involving R, we get

$$R = 3\left[\frac{1}{M_0{}^{(\text{ret})}} + \frac{1}{M_0{}^{(\text{adv})}} \right] \sum_a \int \frac{\delta_4(X, A)}{\sqrt{-g(A)}} \lambda(A)\, da, \tag{96}$$

whence

$$\Box u = \frac{1}{2}\left[\frac{1}{M_0{}^{(\text{ret})}} - \frac{1}{M_0{}^{(\text{adv})}} \right] \sum_a \int \frac{\delta_4(X, A)}{\sqrt{-g(A)}} \lambda(A)\, da, \tag{97}$$

$$\Box v = \frac{1}{2}\left[\frac{1}{M_0^{(\mathrm{adv})}} - \frac{1}{M_0^{(\mathrm{ret})}}\right]\sum_a \int \frac{\delta_4(X, A)}{\sqrt{-g(A)}}\lambda(A)\, da. \tag{98}$$

To obtain the constant A, it is sufficient to solve (97) in the Euclidean approximation, giving

$$A = \frac{N\lambda}{8\pi}\left[\frac{1}{M_0^{(\mathrm{ret})}} - \frac{1}{M_0^{(\mathrm{adv})}}\right], \tag{99}$$

where N is the number of particles in the Sun. Remembering that

$$8\pi G^* = \frac{3}{K} = \frac{6}{M_0^{(\mathrm{ret})} M_0^{(\mathrm{adv})}}, \tag{100}$$

we get

$$A = \frac{\lambda}{6} NG^*[M_0^{(\mathrm{adv})} - M_0^{(\mathrm{ret})}]. \tag{101}$$

Putting

$$m_0 = \frac{1}{2}\lambda[M_0^{(\mathrm{ret})} + M_0^{(\mathrm{adv})}], \tag{102}$$

(101) can be written in the form

$$A = \frac{1}{3} G^* \mathcal{M} \chi, \tag{103}$$

where

$$\chi = \frac{M_0^{(\mathrm{adv})} - M_0^{(\mathrm{ret})}}{M_0^{(\mathrm{adv})} + M_0^{(\mathrm{ret})}}, \qquad \mathcal{M} = Nm_0. \tag{104}$$

Neglecting small terms, $\mathcal{M}$ is the inertial mass of the Sun.

From the 44 component of the gravitational equations, we have

$$\begin{aligned} -\frac{g'}{r} + \frac{1 - e^g}{r^2} &= -\frac{3}{K} e^{f-g}(T_{44} + \Phi_{44}) \\ &\simeq -\frac{3}{K} T_{44} \\ &\simeq -8\pi G^* m_0 \sum_a \int \frac{\delta_4(X, A)}{\sqrt{-g(A)}}, \end{aligned} \tag{105}$$

which gives

$$S = 2G^*\mathfrak{M}. \tag{106}$$

Hence

$$A = \chi S/6$$

and

$$u = 1 + \frac{1}{6}\frac{\chi S}{r} + \frac{D}{r^2} + \dots, \tag{107}$$

$$v = 1 - \frac{1}{6}\frac{\chi S}{r} + \frac{E}{r^2} + \dots, \tag{108}$$

in the exterior. To determine D, E, we note that in the exterior

$$\Box u = \Box v = -\frac{1}{6} R + O\left(\frac{1}{r^5}\right), \tag{109}$$

whence to order r^{-4} we have

$$\frac{d}{dr}\left(r^2 e^{(f-g)/2}\frac{du}{dr}\right) = \frac{d}{dr}\left(r^2 e^{(f-g)/2}\frac{dv}{dr}\right). \tag{110}$$

Substituting for e^f, e^g, from (82) and (83), using (107) and (108) in (110), and equating the coefficients of $1/r$ in the parentheses, we get

$$D - E = \frac{1}{6}\chi S^2. \tag{111}$$

The conformal requirement that K be independent of r imposes the condition $uv = 1$, giving

$$D + E = \frac{1}{36}\chi^2 S^2, \tag{112}$$

so that

$$D = \frac{1}{12}\chi S^2 + \frac{1}{72}\chi^2 S^2, \qquad E = -\frac{1}{12}\chi S^2 + \frac{1}{72}\chi^2 S^2. \tag{113}$$

It will be recalled that our aim in the above work was to consider planetary orbits, given by (86), with respect to the line-element (71) with e^f and e^g of the form (72), S being given by (106). For the line element (71), (72), there would be no difficulty if the inertial mass appearing in (86) were a constant. The orbits would then be geodesic, and there would be no difference from the Einstein theory. But for $\chi^2 \neq 0$ the inertial mass m is not a constant. The above results allow us to determine the form of m, starting with

$$m = \frac{1}{2}\lambda[M^{(\text{ret})} + M^{(\text{adv})}]. \tag{114}$$

From (92), (107), (108), (113), we have

$$M^{(\text{ret})} = M_0^{(\text{ret})}\left(1 + \frac{1}{6}\frac{\chi S}{r} + \frac{1}{12}\frac{\chi S^2}{r^2} + \frac{1}{72}\frac{\chi^2 S^2}{r^2} + \ldots\right), \tag{115}$$

$$M^{(\text{adv})} = M_0^{(\text{adv})}\left(1 - \frac{1}{6}\frac{\chi S}{r} - \frac{1}{12}\frac{\chi S^2}{r^2} + \frac{1}{72}\frac{\chi^2 S^2}{r^2} + \ldots\right). \tag{116}$$

Substituting (115), (116), in (114), and using the definition

$$m_0 = \frac{1}{2}\lambda[M_0^{(\text{ret})} + M_0^{(\text{adv})}], \tag{117}$$

we obtain, after an easy reduction,

$$m = m_0\left(1 - \frac{1}{6}\frac{\chi^2 S}{r} - \frac{5}{72}\frac{\chi^2 S^2}{r^2} + \ldots\right). \tag{118}$$

Next we recall that the model under investigation, with variable gravity according to (70), arises because either $M_0^{(\text{ret})} \gg M_0^{(\text{adv})}$, with (65) holding for $M_0^{(\text{ret})}$, $M_0^{(\text{adv})}$, or $M_0^{(\text{ret})} \ll M_0^{(\text{adv})}$ with (66) holding. In either case $\chi^2 \simeq 1$, as can be seen from (104), so that

$$m \simeq m_0\left(1 - \frac{1}{6}\frac{S}{r} - \frac{5}{72}\frac{S^2}{r^2} + \ldots\right). \tag{119}$$

It will be shown below that (119) gives $\frac{3}{4}$ of the Einstein value for the bending of light, and $\frac{2}{3}$ of the Einstein value for the rotation of the perihelion of Mercury. Only if $\chi^2 = 0$, which requires $M_0^{(\text{ret})} = M_0^{(\text{adv})}$ do we get the Einstein values. In §9.5 we found that the variable-gravity model evolves with increasing cosmic time into the Einstein–de Sitter model. In the latter we do have $M_0^{(\text{ret})} = M_0^{(\text{adv})}$, and hence the Einstein values. During the transition from one model to another, χ^2 changes from unity to zero. Correspondingly, there is a change for the bending of light and for the perihelion rotation from values appreciably less than in the Einstein theory to the same values as in the Einstein theory.

We turn now to the effect of (119) on the orbital equations (86). By writing

$$m(A)\frac{d}{da} = m_0\frac{d}{d\tilde{a}}, \tag{120}$$

we can express (86) in the form

$$\frac{d^2a^i}{d\tilde{a}^2} + \Gamma^i{}_{kl}\frac{da^k}{d\tilde{a}}\frac{da^l}{d\tilde{a}} = \frac{mm^i}{{m_0}^2}. \tag{121}$$

Since $m^i = 0$, $i = 2, 3, 4$, the equations for these components are the same as in Einstein's theory, except that $\tilde{a}$ replaces a. There are similar integrals of motion (Eddington, 1930, p. 86),

$$r^2\frac{d\phi}{d\tilde{a}} = h, \qquad \frac{dt}{d\tilde{a}} = ce^{-f}. \tag{122}$$

The line element gives

$$e^g\left(\frac{dr}{d\tilde{a}}\right)^2 + r^2\left(\frac{d\phi}{d\tilde{a}}\right) - e^f\left(\frac{dt}{d\tilde{a}}\right)^2 = -\left(\frac{da}{d\tilde{a}}\right)^2 = -\left(\frac{m}{m_0}\right)^2, \tag{123}$$

the orbit being taken to lie in the plane $\theta = \pi/2$. To within the post-Newtonian approximation, we can write $e^g = e^{-f}$ in (123), in which case the situation in (123) is as in Einstein's theory except that $-(m/m_0)^2$ appears on the righthand side instead of -1. Following a treatment similar to that given by Eddington, we write $u = 1/r$ and use (122) to express (123) in the form

$$\left(\frac{du}{d\phi}\right)^2 + u^2 = \frac{c^2}{h^2} + Su^3 - \frac{e^f}{h^2}\left(\frac{m}{m_0}\right)^2. \tag{124}$$

Substituting for m/m_0 from (119), we now get

$$\left(\frac{du}{d\phi}\right)^2 + u^2 = \frac{c^2 - 1}{h^2} + \frac{2k}{h^2}u - \frac{k^2}{2h^2}u^2 + \frac{3}{2}ku^3 + O(u^4), \tag{125}$$

where

$$k = \frac{2}{3}S. \tag{126}$$

Differentiating (125) with respect to ϕ, and canceling $du/d\phi$ throughout, gives

$$\frac{d^2u}{d\phi^2} + u = \frac{k}{h^2} - \frac{k^2}{2h^2}u + \frac{9}{4}ku^2 + \cdots. \tag{127}$$

From the first term on the righthand side of (127), we see that k takes the place that $\frac{1}{2}S$ holds in the usual theory. Therefore it is k, not S, that is determined by the observed planetary motions. But it is S, not k, that determines the bending of electromagnetic radiation by the Sun. Hence we have

$$k = \frac{2}{3}S = \text{observationally determined quantity}, \tag{128}$$

whereas in the usual theory

$$\frac{1}{2}S = \text{same observationally determined quantity}. \tag{129}$$

Evidently, S in the present theory is $\frac{3}{4}$ of the usual S, and hence the bending effect is only $\frac{3}{4}$ of its usual value.

For the perihelion motion of a planet of eccentricity e, the first-order solution of (127) can be written in the form

$$u = \frac{k}{h^2}[1 + e\cos(\phi + \omega)]. \tag{130}$$

Substituting (130) for u in the post-Newtonian terms of (127), we get

$$\frac{d^2u}{d\phi^2} + u = \frac{k}{h^2} - \frac{k^3}{2h^4}[1 + e\cos(\phi + \omega)] + \frac{9k^3}{4h^4}[1 + e\cos(\phi + \omega)]^2 + \cdots. \tag{131}$$

The coefficient of $\cos(\phi + \omega)$ on the righthand side of (131) is $4ek^3/h^4$, whereas in the Einstein theory the corresponding coefficient is $6ek^3/h^4$. Since this term gives rise to the perihelion motion, the present theory is seen to give a motion equal to $\frac{2}{3}$ of the Einstein value.

The possibility that observations of the bending of light might lead to a result less than the Einstein value has been considered by Dicke, and Dicke and Goldenberg (1967) have also suggested that if the interior of the Sun is rotating comparatively rapidly, the relativistic rotation of the perihelion of Mercury would need to be less than the Einstein value. However, the data, while somewhat equivocal, appear to be against this point of view (Clemence, 1947; Muhleman *et al.*, 1970; Sramek, 1971). Since the above deviations from the Einstein theory are more severe than those considered by Dicke, we must conclude that existing data make the variable-gravity theory based on equations (63) to (70) untenable—at any rate given the present value of the cosmic time τ. Since the variable-gravity model evolves with increasing cosmic time into the Einstein–de Sitter model, for which the classical tests hold good, it would be possible to argue that variable gravity occurred at earlier epochs than the present, i.e., that transition to the Einstein–de Sitter model has by now taken place. But this would be a weak position. It is much preferable to argue that observations of the bending of light and of the perihelion motion show that $M^{(\mathrm{ret})} = M^{(\mathrm{adv})}$, and hence provide strong evidence for the response condition (115) of Chapter 6, which point of view we shall adopt from here on.

§9.7. ELECTROMAGNETIC RESPONSE OF THE UNIVERSE (LECTURE 26)

We turn now to the first of the difficulties set out in §9.1. In the early chapters the electromagnetic theory was developed in terms of a time-symmetric interaction between particles. This theory was shown to lead to results in agreement with experiment, for quantum electrodynamics as well as for classical theory, provided the universe gave an appropriate form of response. However, none of the Friedmann models were satisfactory for meeting the required response condition. We ask now whether this situation has been changed by the considerations of §9.5.

It will be useful to begin an answer to this question by illustrating the difficulty for the Einstein–de Sitter model, doing so with respect to the Minkowski conformal frame instead of to the Robertson-Walker frame as in the early chapters. We have

$$ds^2 = d\tau^2 - dr^2 - r^2(d\theta^2 + \sin^2\theta \, d\varphi^2), \tag{132}$$

$$M^{(\mathrm{ret})} = M^{(\mathrm{adv})} = \frac{1}{2}\lambda L^{-3}\tau^2, \tag{133}$$

$$m = \frac{1}{2}\lambda[M^{(\mathrm{ret})} + M^{(\mathrm{adv})}] = \frac{1}{2}\lambda^2 L^{-3}\tau^2, \tag{134}$$

where L^{-3} is the particle density, which is constant in the Einstein–de Sitter model. Although most of the particles may become locally condensed into galaxies, stars, and perhaps dark objects collapsing relativistically, we suppose that some finite gaseous residue is maintained throughout space at all τ, and we write N for the particle density of this residue.

The exponential attenuation coefficient for an electromagnetic wave of frequency ν propagating in an ionized gas of electron density N is given by

$$\sim(\text{constant})\frac{Ne^2}{m\omega^2}\nu \qquad \text{if } \omega > \nu, \tag{135}$$

and by

$$\sim(\text{constant})\frac{Ne^2}{m\nu} \qquad \text{if } \nu > \omega, \tag{136}$$

where ν is the effective collision frequency. Remembering that ω does not depend on τ in the Minkowski frame, the required response condition is that the integral taken with $\tau \to \infty$ of (135) or (136), whichever is appropriate, shall be divergent. Suppose that ν does not tend to zero or to infinity as $\tau \to \infty$. The behaviour of the integral is then determined by

$$\int^{\infty}\frac{d\tau}{m}, \tag{137}$$

and with $m \propto \tau^2$ this is convergent, not divergent. Nor would the situation be helped by $\nu \to 0$ or by $\nu \to \infty$ with τ, since in either of these cases the integral would be made more convergent.

The difficulty here is due to $m \to \infty$ as $\tau \to \infty$, which is just the behaviour we were enjoined in §9.5 to avoid. It was repeatedly emphasized there that $m \propto \tau^2$ *applies only sufficiently close to a surface of zero mass.* We cannot continue to use (133), (134), at large distance from such a surface. Provided the positive and negative contributions to the mass field prevent $m \to \infty$, just as the plus and minus charge contributions prevent divergences of the electromagnetic potential, the difficulty concerning (137) disappears.

The above considerations, together with those of §9.6, bear critically on the whole problem of response conditions. The situation can be summarized in the following terms:

I. The classic tests of relativity, agreeing with the Einstein values for the bending of light and for the rotation of the perihelion of Mercury, provide strong evidence for the existence of a mass-response condition, of the form given in (115) of Chapter 6.

II. The need for the electromagnetic-response condition to be satisfied implies that the universe does not possess the simple structure of a Friedmann model. There is strong support for positive and negative contributions to the mass field, and for the view that the region of spacetime so far investigated by astronomers is but a small part of a far vaster structure.

§9.8. LENGTH SCALES AND UNITS

In §6.1 we saw how all of physics, excluding gravitation, can be described by:

(1) three dimensionless coupling constants, which are the fine-structure constant and constants for the strong and weak interactions;
(2) the particle mass ratios; and
(3) the mass of some chosen reference particle.

It is (3) which introduces dimensionality into physics. The issue now is whether the inclusion of gravitation and cosmology throws further light on the previous situation. In particular, since we have developed a theory of the nature of mass, can we obtain insight into (3)? We might hope, for example, to determine the mass of a reference particle in terms of the cosmological structure.

In the following discussion we shall continue, for simplicity, to use the Einstein–de Sitter cosmological model. First we note that the definition

$$8\pi G = \frac{6\lambda^2}{m^2}, \tag{138}$$

with m given by (134), leads to a well-known relation between the cosmological mass density $\sigma = mL^{-3}$ and the Hubble constant H. In the Minkowski conformal frame, H and τ are related by

$$H = 2/\tau. \tag{139}$$

These relations lead easily to

$$\sigma = \frac{3H^2}{8\pi G}. \tag{140}$$

It follows from this derivation that (140) was contained within the previous discussion, and is indeed no more than an expression of the definition of G. We see therefore that (140) cannot serve to improve our understanding of the issues set out above.

We cannot hope to determine m from the gravitational theory alone, because the gravitational theory does not depend on the numerical value of λ^2, whereas m is so dependent. The value of λ^2 becomes important, however, when the gravitational theory is combined with electrodynamics. Writing (134) in the form

$$\lambda^2 = \frac{1}{2}(LH)^3 \cdot (mH^{-1}), \tag{141}$$

we have λ^2 expressed as the product of two dimensionless numbers. $(LH)^3$ is the reciprocal of the number of particles within distance $\sim H^{-1}$, whereas mH^{-1} is the ratio of the scale of the observed portion of the universe to the Compton wavelength of our reference particle. Empirically, we have the following known values

$$(LH)^3 \simeq 10^{-79}, \tag{142}$$

$$mH^{-1} \simeq 10^{39}. \tag{143}$$

The number in (142) arises within the gravitational theory alone, but the determination of (143) requires electrodynamics as well as gravitation. The number in (143) refers to the electron as reference particle. Other choices for the reference particle would not change the fact that the combination

$$(mH^{-1})^2 \cdot (LH)^3 \tag{144}$$

is of order unity. Nor would the general order of magnitude of λ^2 be changed,

$$\lambda^2 \simeq 10^{-40}. \tag{145}$$

How are we to interpret (145)? We could take the point of view that the emergence of a dimensionless number as small as 10^{-40} is nothing new. In the usual theory, G is introduced as an *ad hoc* constant (see §6.5) with the dimensionality of $(\text{mass})^{-2}$, such that when multiplied by the square of a typical particle mass, for example, the experimentally determined mass m of the electron, it gives $Gm^2 \simeq 10^{-40}$. Since in the present theory G is related to λ^2 by (138), one can seek to argue that our finding $\lambda^2 \simeq 10^{-40}$ should be no surprise. This point of view hardly seems satisfactory, however, because in the present theory λ is the basic constant, not G [the relation (138) only defines G in terms of λ], and it seems less than plausible to suppose that a basic physical coupling constant could be as small as 10^{-20}.

Our point of view is that the emergence of (145) is a clear indication of the incompleteness of the theory. Much remains to be discovered, but the road towards success probably involves the structure of the particles themselves. The relation of modern particle physics to gravitation and cosmology is still unexplored territory, but it is in this territory where further insight into (145) is likely to be found. Such considerations lie, however, beyond the scope of this book.

10

SUMMARY

We have been concerned throughout this book with the thesis that the behavior of physical systems is a compound of certain basic laws and of an interaction with the universe in the large. The basic laws are more simply stated than in the usual field-theory formulation of physics. We have found considerable similarity in the responses of the universe to different kinds of interaction. In Chapter 4 we formulated the response condition for a classical electromagnetic field in terms of a spacetime slab $0 \leq t \leq T$. Writing

$$B^i{}_{t<0} + B^i{}_{t>T} \tag{1}$$

for the part of the field that could not be attributed to direct-particle interactions within the slab, we obtained

$$B^i{}_{t>T}(X) = [B^i(X)_{\text{ret}} - B^i(X)_{\text{adv}}]_{0 \leq t \leq T} \tag{2}$$

Explicitly for the righthand side of (2), in (117) of Chapter 4 we had

$$[B^i(X)_{\text{ret}}]_{0 \leq t \leq T} = \frac{1}{2} \sum_b A^{i(b)}(X)_{\text{ret}}, \tag{3}$$

$$[B^i(X)_{\text{adv}}]_{0 \leq t \leq T} = \frac{1}{2} \sum_b A^{i(b)}(X)_{\text{adv}}, \tag{4}$$

$A^{i(b)}{}_{\text{ret}}$, $A^{i(b)}{}_{\text{adv}}$, being the fully retarded and the fully advanced potentials of particle b, obtained from the segment of the path of b lying within the slab.

A similar formulation for mass fields, in (115) of Chapter 6, gave

$$M_{t>T} = M^{(\text{ret})}{}_{0 \leq t \leq T} - M^{(\text{adv})}{}_{0 \leq t \leq T}. \tag{5}$$

In both cases the response from the future is expressed as a difference between the retarded and advanced interactions within the slab.

The quantum form of (2), given in (124) of Chapter 4, was

$$B_i(X)_{t>T} = \frac{1}{2} \sum_b \{[A_i^{(b)}(X)_{\text{ret}+} - A_i^{(b)}(X)_{\text{adv}+}] + [A_i'^{(b)}(X)_{\text{ret}-} - A_i'^{(b)}(X)_{\text{adv}-}]\}, \tag{6}$$

where $A_i^{(b)}$, $A_i'^{(b)}$, were calculated for paths and conjugate paths, respectively. When paths and conjugate paths coincide—as they do in the classical case—the form (6) becomes the same as (2). In (26) of Chapter 8 we had a response from the future for the free-particle propagator,

$$\frac{1}{2}[(K_{\text{ret}+} - K_{\text{adv}+}) + (K'_{\text{ret}-} - K'_{\text{adv}-})], \tag{7}$$

in strikingly similarity to (6). Since the quantum response (6) contains subtleties not present in the classical response (2), it seems likely that (5) requires extension to a more far-reaching form, possibly involving the structure of particles.

It is noteworthy that, although all the results of quantum electrodynamics follow from (6), the appearance of both paths and conjugate paths implies that there is no purely amplitude formulation of the theory. Probabilities, not amplitudes, are necessarily involved, so that the redundancy present in the usual theory does not appear here.

The influence functionals given in (127) and (129) of Chapter 6 show a clear distinction between scattering problems and decay problems. The normal procedure in scattering is to add amplitudes for different frequencies before taking the square of the modules to give a probability. In decay problems, on the other hand, the square of the modulus for each separate frequency is taken first, and the resulting moduli are then added for the total probability. This procedure is necessary to give an appropriate time-dependence for the decay of a quantum state. It is usually argued that, whereas the frequencies for scattering are determined by the δ_+ function and are therefore systematically phased with respect to each other, the frequencies in decay problems arise from randomly phased field oscillators. The total amplitude should always be obtained by adding all frequencies, but in decay problems cross-products between different frequencies average to zero when the square of the modulus is taken. No doubt a consistent theory can be formulated in this way, but the issue is treated scantily if at all in most quantum texts; so the issue would seem to cause some embarrassment.

The situation is clear-cut in the present theory. The influence functionals contain couplings of paths with paths, of conjugate paths with conjugate paths, and of paths with conjugate paths. From (127) and (129) of Chapter 4, it is seen that when a path interacts with a conjugate path it is the same frequency that operates on both. There is no case in which a frequency K interacting with a path is combined with a different frequency K' acting on a conjugate path. This is the decay problem. When paths interact with paths, on the other hand, the interaction frequencies have no correlation with those that operate in the coupling of conjugate paths with conjugate paths. This is the scattering case.

We began the program of work set out in this book with the initial aim of obtaining a quantum generalization of (2). It gradually became clear, however, that a commitment to the electrodynamic case must lead to extensions of the action-at-a-distance point of view to all parts of physics. So far we have obtained similar results in (5) and (7). But other major issues remain, particularly for those parts of physics that involve short-range forces. One may wonder how a point of view that has been based on long-range forces could come to be applied to parts of physics requiring short-range forces. How this may be done has been considered elsewhere (F. Hoyle and J. V. Narlikar, *Il Nuovo Cimento,* 7A, 262, 1972).

In the course of the work, we were led to the view that physical theories should be conformally invariant. The development of a conformally invariant theory of gravitation freed us from the need to work always in a conformal frame in which particle masses are constants. It is the use of such a frame that leads to metric singularities, for example, at the so-called origin of the universe or at the implosion of a black hole. With particles acquiring their mass by interaction with other particles, we are led to realize that there could be surfaces where the particle masses are zero. Whereas the conformally invariant theory experiences no difficulty with such surfaces, the usual theory encounters metric singularities which prevent proper continuity across them from being established. The usual theory erects artificial boundaries at these surfaces, limiting the universe to only one side of an open surface, or to the interior of a closed surface.

These considerations led us to the view that the universe is probably a much larger structure than it is usually supposed to be, a conclusion which may well form a base for new cosmological developments.

APPENDIXES

To avoid overloading the main text with subsidiary detail, we have relegated several investigations to these appendixes. Each appendix is keyed to the section to which it is, in effect, a note. However, before we deal with these necessary additions, it may be useful to consider the topic of generalized functions. In our experience it is fairly common for students to continue thinking in terms of conventional analysis while carrying through operations with the Dirac delta function. Such operations are then uneasily executed —facility depends, of necessity, on suitably broadening the concept of "function."

APPENDIX 1. GENERALIZED FUNCTIONS

The Dirac delta function belongs to a class referred to in the literature sometimes as generalized functions, sometimes as ideal functions, and sometimes as "distributions." These entities are not functions in the sense of elementary mathematics, as will be clear from the following definition of the delta function:

$$\delta(x) = 0 \text{ for } x \neq 0, \qquad \int_{-\infty}^{\infty} \delta(x)\,dx = 1. \tag{1}$$

Here $\delta(x)$ is well-defined for all points on the real axis except $x = 0$. The second part of the definition includes integration through $x = 0$. For such an integration to be applied to an ordinary function, we would need to know its value at $x = 0$. Yet this information is deliberately withheld in (1). In going from ordinary to generalized functions, we drop some of the rules governing the former while still retaining certain basic operations. Generalized functions can be defined in several apparently different ways, although it can be proved that the different approaches are actually equivalent to one another. It will be sufficient here to consider two of these approaches, which can usefully be illustrated in terms of the delta function. A third, more abstract method is given by Courant and Hilbert (1962, II, 774), where the equivalence of the three methods is proved.

1. Consider a finite domain θ with respect to a real variable x, and let the set of test functions $\phi(x)$ be zero outside θ. If all derivatives $\phi^{(r')}$, $r' \leq r$, are continuous, but $\phi^{(r+1)}$ is not, we say that ϕ is of class C^r. If $\phi^{(r)}$ is continuous for all r, we say that ϕ is of class C^∞. Unless otherwise stated, ϕ will be taken to be of class C^∞.

Let $\mathfrak{L}$ be a linear differential operator of order r having coefficients that are differentiable at least r times within a closed subdomain $\bar{\theta}$ of θ. Define an adjoint operator $\mathfrak{L}^*$ by

$$(\psi, \phi) = \int_{\bar{\theta}} \mathfrak{L}[\psi]\phi \, dx = \int_{\bar{\theta}} \psi \mathfrak{L}^*[\phi] \, dx, \tag{2}$$

where ψ, ϕ, are test functions at least of class C^r chosen to vanish outside $\bar{\theta}$. Now let W be any continuous (or piecewise continuous) function in $\bar{\theta}$. We make no statement concerning the differentiability of W in the usual analytic sense. Yet we define

$$f = \mathfrak{L}[W] \tag{3}$$

to be an ideal function with the meaning that

$$(f, \phi) = \int_{\bar{\theta}} \mathfrak{L}[W]\phi \, dx = \int_{\bar{\theta}} W\mathfrak{L}^*[\phi] \, dx \tag{4}$$

for any test function $\phi(x)$ at least of class C^r chosen to vanish outside $\bar{\theta}$. The righthand side of (4) is well-defined in the usual sense.

The delta function $\delta(x)$ can be defined according to (4) by taking

$$\mathfrak{L} \equiv \frac{d}{dx}, \qquad W = 1 \text{ for } x \geq 0, \qquad W = 0 \text{ for } x < 0, \tag{5}$$

with $\bar{\theta}$ any part of the real axis including $x = 0$. It is easy to verify that for any ϕ which vanishes on the boundary of $\bar{\theta}$ and which is differentiable at least once,

$$\int_{\bar{\theta}} \delta(x)\,\phi(x)\,dx = \phi(0). \tag{6}$$

To define $\delta'(x)$, we choose $\mathfrak{L}$ to be d^2/dx^2, take the same W, and use test functions ϕ that are differentiable at least twice and that are zero outside $\bar{\theta}$. Then we have

$$\int_{\bar{\theta}} \delta'(x)\,\phi(x)\,dx = \int_{\bar{\theta}} W(x)\frac{d^2\phi}{dx^2}\,dx = -\phi'(0). \tag{7}$$

2. Suppose we have a sequence of functions $f_n(x)$ continuous in θ. Let

$$\lim_{\substack{n\to\infty \\ m\to\infty}} \int_{\bar{\theta}} |f_n(x) - f_m(x)|\,\phi(x)\,dx = 0, \tag{8}$$

and let the limiting process be uniform with respect to all test functions $\phi(x)$ with the property that $|\phi(x)|$, $|\phi^{(r')}(x)|$, $r' \leq r$, are bounded by the same number M in the closed subdomain $\bar{\theta}$ of θ. The sequence $[f_n(x)]$ is then said to be weakly r-convergent in θ.

We now define a generalized function $f(x)$,

$$f(x) = \lim_{n\to\infty} f_n(x), \tag{9}$$

with the meaning that

$$\int_{\bar{\theta}} f(x)\,\phi(x)\,dx = \lim_{n\to\infty} \int_{\bar{\theta}} f_n(x)\,\phi(x)\,dx. \tag{10}$$

The limit on the righthand side of (10) is always well-defined in the usual analytic sense in virtue of (8). The righthand side of (9), on the other hand, may not exist in the usual sense.

To illustrate the case of the delta-function, consider

$$f_n(x) = \frac{n}{\sqrt{\pi}}\,e^{-n^2x^2}, \qquad -N \leq x \leq N, \tag{11}$$

where n is taken large. In this case (8) holds for all test functions ϕ with $|\phi(x)| \leq M$ throughout $\bar{\theta}$, i.e., for $r = 0$, where $\bar{\theta}$ is any closed interval within $[-N, N]$. Evidently

$$\lim_{n\to\infty} \int_{\bar{\theta}} f_n(x)\, \phi(x)\, dx = \phi(0) \tag{12}$$

provided $\bar{\theta}$ contains $x = 0$, from which it follows that

$$\lim_{n\to\infty} f_n(x) = \delta(x). \tag{13}$$

Although in the usual analytic sense the functions $f_n(x)$ become unbounded at $x = 0$ as $n \to \infty$, we can define (13) in terms of (10).

It is to be emphasized that a generalized function does not have meaning by itself, but only in relation to θ, $\bar{\theta}$, and to the specification of the test functions. For a given sequence $[f_n(x)]$, in order that (8) shall continue to hold when $\bar{\theta}$ is changed, it may be necessary to alter the specification of the test functions—extending $\bar{\theta}$ may require the value of r to be increased, for example. Hence the choice of suitable test functions may be related to the interval over which a generalized function is defined.

We illustrate the difference between generalized functions and ordinary functions with another example. Let $\bar{\theta}$ be an interval containing $x = 0$, and let $\phi(x)$ be a test function that can be differentiated twice throughout $\bar{\theta}$ and which is zero outside θ. Relative to such functions and to $\bar{\theta}$, we can write

$$x\delta'(x) = -\delta(x), \tag{14}$$

since both sides of (14) when integrated over $\bar{\theta}$ against any such $\phi(x)$ give $-\phi(0)$. Suppose, however, that we attempt to divide both sides of (14) by x,

$$\delta'(x) = -\frac{\delta(x)}{x}. \tag{15}$$

The lefthand side of (15) when integrated over $\bar{\theta}$ against any ϕ gives $-\phi'(0)$ and so is well-defined, but in order that the righthand side be also well-defined, it would be necessary for $\phi(x)/x$ to be differentiable throughout $\bar{\theta}$ at least once. This need not be the case. For example, take the interval $[-1, 1]$ for $\bar{\theta}$ and $(1 - x^2)^n$, $n > 1$, for $\phi(x)$. This satisfies the requirement for a test function applicable to (14) or to the lefthand side of (15), but not to the righthand side of (15), since $\phi(x)/x$ is not differentiable at $x = 0$.

Generalized functions appear in physics through propagators which satisfy wave equations of the second order. In the theory developed in the main text, these propagators describe interactions between pairs of particles. Characteristically in the direct-particle theory, the propagators are subject to integration at each member of a particle pair. From a mathematical point of view, the propagators have no existence in themselves—they exist only in respect of test functions and domains of integration determined by the particles. Yet in conventional field theory the same generalized functions are used in respect of fields which are taken to have an existence independent of the particles. From what has just been said, it will be clear that such a procedure is mathematically dubious. Evidently, we have a cogent argument favoring the direct-particle theory against conventional field theory. We end this discussion by considering some generalized functions which are important in their relation to physical propagators.

Consider two points $x^i = 0$, $x^i = (\boldsymbol{x}, t)$, in Minkowski space, and write

$$r^2 = \boldsymbol{x} \cdot \boldsymbol{x}, \qquad s^2 = t^2 - r^2. \tag{16}$$

For the generalized function $\delta(s^2)$, we have

$$\delta(s^2) \equiv \delta(t^2 - r^2) = \frac{1}{2r}[\delta(t - r) + \delta(t + r)] \tag{17}$$

with respect to test functions of x^i having nonsingular first derivatives. The well-known Dirac identity

$$\Box_x \delta(s^2) = 4\pi \, \delta(\boldsymbol{x}) \, \delta(t) \tag{18}$$

can be proved in the following way. First we notice that any spherically symmetric solution of $\Box f = 0$ is expressible for $r \neq 0$ as a linear combination

$$\frac{1}{r} g(t - r) + \frac{1}{r} h(t + r), \tag{19}$$

where g, h, are functions of $t - r$ and $t + r$, respectively. The righthand side of (17) is of this form, so that $\Box_x \delta(s^2) = 0$ except possibly at the point $r = 0$, $t = 0$. To prove that $\Box_x \delta(s^2)$ is a generalized function equal to the righthand side of (18), with respect to the test functions described above we have to integrate over a closed domain in spacetime that includes $x^i = 0$. Taking the righthand side of (17) to represent $\delta(s^2)$ and considering, first, the term

$$\frac{\delta(t - r)}{2r} = f, \text{ say}, \tag{20}$$

and letting the domain be a 4-cylinder whose cross section in space is the sphere $r \leq R$ and whose generators are between $-T \leq t \leq T$, $T \geq R$, we get

$$\int \Box_x f \, d^4x = \int f_i \, dS^i$$

$$= \int_0^R \frac{\partial f}{\partial t}\bigg|_{t=T} 4\pi r^2 \, dr - \int_{-T}^{T} \frac{\partial f}{\partial r}\bigg|_{r=R} 4\pi R^2 \, dt$$

$$= \int_0^R \frac{\delta'(T-r)}{2r} 4\pi r^2 \, dr + \int_{-T}^{T} \left[\frac{\delta'(t-R)}{2R} + \frac{\delta(t-R)}{2R^2}\right] 4\pi R^2 \, dt$$

$$= -2\pi R\, \delta(T-R) + 2\pi R\, \delta(T-R) + 2\pi = 2\pi. \tag{21}$$

In the second line here we have made use of the fact that $f = 0$ at $t = -T$. Hence there is no surface integral at $t = -T$. A similar calculation for the second term on the righthand side of (17) also gives 2π. Hence,

$$\int \Box_x \, \delta(s^2) \, d^4x = 4\pi, \tag{22}$$

the same as the integral of the righthand side of (18), which is therefore established.

By using (18), one can derive the Fourier representation of $\delta(s^2)$ very quickly. Write

$$\delta(s^2) = \int \lambda(k^i) \exp(-ik \cdot x) \frac{d^4k}{(2\pi)^4}, \tag{23}$$

where $k \cdot x = k_i x^i$. Since

$$\delta(\boldsymbol{x})\, \delta(t) = \int \exp(-ik \cdot x) \frac{d^4k}{(2\pi)^4}, \tag{24}$$

it is easy to see that

$$\lambda(k^i) = -\frac{4\pi}{k^2}. \tag{25}$$

[To prove that the righthand side of (24) is the same generalized function as the lefthand side, with respect to a spacetime domain including $x^i = 0$ and to suitable test functions, integrate each component of the vector k^i over the interval $[-K, K]$. By taking an increasing sequence of values of K, one obtains a sequence of functions which tends in the limit $K \to \infty$ to the lefthand side of (24). This is easily seen by using the second of the two ways discussed above for defining generalized functions.]

Care is needed when (25) is substituted in (23) to take the principal part of the resulting integral,

$$\delta(s^2) = -\int 4\pi \frac{\mathscr{P}}{k^2} \exp(-ik \cdot x) \frac{d^4k}{(2\pi)^4}. \tag{26}$$

The need to take the principal part arises at the poles of $1/k^2$. Writing $k^i = (\boldsymbol{k}, \omega)$, we have poles in a complex ω-plane at

$$\omega = \pm|\boldsymbol{k}|. \tag{27}$$

If instead of taking the principal part at each pole we were to take the residue at $\omega = +|\boldsymbol{k}|$ only, we would get $\delta_+(s^2)$. This is often stated in the form

$$\delta_+(s^2) = -\int \frac{4\pi}{k^2 + i\varepsilon} \exp(-ik \cdot x) \frac{d^4k}{(2\pi)^4}. \tag{28}$$

Other solutions of $\Box_x f = 4\pi\, \delta(\boldsymbol{x})\, \delta(t)$, e.g., $\delta_-(s^2)$, $\delta(t - r)/r$, $\delta(t + r)/r$, can be obtained by choosing suitable contours in the ω-plane. These are shown in Figure A1.1.

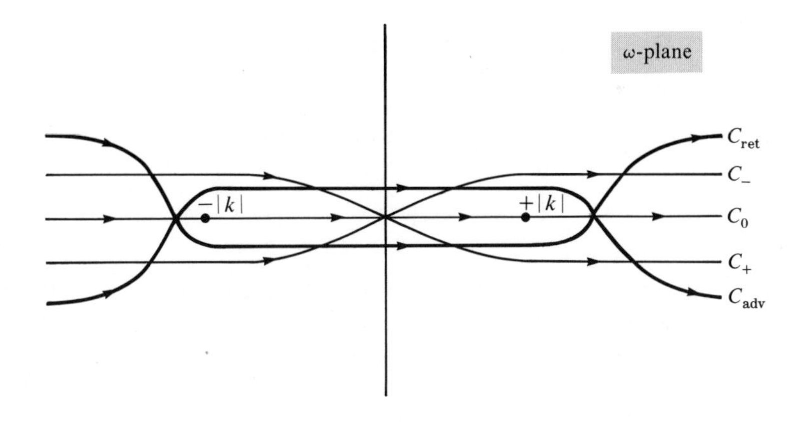

Contour	Solution
C_0	$\delta(S^2)$
C_+	$\delta_+(S^2)$
C_-	$\delta_-(S^2)$
C_{ret}	Full retarded solution
C_{adv}	Full retarded solution

FIGURE A1.1
Solutions of $f = 4\pi\delta(\mathbf{x})\ \delta(t)$.

APPENDIX 2 (a note to §4.2)

In the nonrelativistic theory the propagator $K(2; 1)$ is given by the path integral,

$$K(2; 1) = \int \exp\left\{\frac{i}{\hbar} S[\Gamma_{21}]\right\} \mathcal{D}\Gamma_{21}, \tag{29}$$

for $t_2 > t_1$, whereas for $t_2 < t_1$ we have $K(2; 1) = 0$. Given the wave function $\psi(\boldsymbol{x}_1, t_1)$ on $t = t_1$,

$$\psi(\boldsymbol{x}_2, t_2) = \int K(2; 1)\psi(\boldsymbol{x}_1, t_1)\, d^3\boldsymbol{x}_1, \qquad t_2 > t_1. \tag{30}$$

The connection with the main text is contained in the question: what differential equation does $\psi(\boldsymbol{x}_2, t_2)$ satisfy? We answer this question with the help of a specific example.

Consider a particle of mass m moving in one direction, the x-direction, say. Let $V(x, t)$ be the potential of the particle in some external field. Then for a given path Γ_{21} we have

$$S[\Gamma_{21}] = \int_{t_1}^{t_2} \left[\frac{1}{2}m\dot{x}^2 - V(x, t)\right] dt. \tag{31}$$

Writing $t_2 = t_1 + \varepsilon$, we can express (30) in the form

$$\psi(x, t_1 + \varepsilon) = \int_{-\infty}^{\infty} K(x, t_1 + \varepsilon; x - \eta, t_1)\psi(x - \eta, t_1)\, d\eta. \tag{32}$$

Taking ε small, all nonrelativistic velocities $\dot{x}$ for the particle are contained in a range of η of the order ε. Hence to the first order in small quantities, the contribution of the potential term in (31) is $-\varepsilon\, V(x, t_1)$. To this order (29) gives

$$K(2; 1) = \left[1 - \frac{i\varepsilon}{\hbar} V(x, t_1)\right] \int \exp\left(\frac{1}{2}\frac{im}{\hbar}\int_{\Gamma_{21}} \dot{x}^2\, dt\right) \mathcal{D}\Gamma_{21}. \tag{33}$$

The path integral in (33) is just the free-particle propagator, the evaluation of which forms the basis of calculations using the path-integral method (Feynman and Hibbs, 1965, p. 42). In the present case this evaluation gives

$$K(x, t_1 + \varepsilon; x - \eta, t_1) = \left(\frac{m}{2\pi i\varepsilon\hbar}\right)^{1/2} \left[1 - \frac{i\varepsilon}{\hbar} V(x, t_1)\right] \exp\left(\frac{im\eta^2}{2\varepsilon\hbar}\right). \tag{34}$$

Expanding the wave functions in (32),

$$\psi(x - \eta, t_1) = \psi(x, t_1) - \eta \frac{\partial \psi}{\partial x} + \frac{1}{2}\eta^2 \frac{\partial^2 \psi}{\partial x^2} + \dots, \tag{35}$$

$$\psi(x, t_1 + \varepsilon) = \psi(x, t_1) + \varepsilon \frac{\partial \psi}{\partial t} + \dots, \tag{36}$$

we get

$$\psi + \varepsilon \frac{\partial \psi}{\partial t} + \dots = \left(\frac{m}{2\pi i \varepsilon \hbar}\right)^{1/2} \left(1 - \frac{i\varepsilon}{\hbar} V\right) \int_{-\infty}^{\infty} \exp\left(\frac{im\eta^2}{2\hbar\varepsilon}\right) \left(\psi - \eta \frac{\partial \psi}{\partial x} + \frac{1}{2}\eta^2 \frac{\partial^2 \psi}{\partial x^2} + \dots\right) d\eta, \tag{37}$$

where ψ and its derivatives and V are taken at x, t_1. For convenience in evaluating the righthand side of (37), the range of η is taken from $-\infty$ to ∞. This is permitted because of the convergence produced by the rapidly oscillating exponential. In lowest order (37) is satisfied identically. The term of first order in η evidently vanishes, and integration of the η^2 term yields a quantity of order ε, which, on being equated to the ε term on the lefthand side, gives

$$\frac{\partial \psi}{\partial t} = -\frac{i}{\hbar} V\psi + \frac{i\hbar}{2m} \frac{\partial^2 \psi}{\partial x^2}. \tag{38}$$

This is the one-dimensional Schrödinger equation. The same argument can readily be extended to obtain the three-dimensional Schrödinger equation.

APPENDIX 3 (a note to §4.7)[1]

§A3.1. A Path-Integral Method Leading to the Perturbation Expansion in Terms of the K_+ Propagator

We begin with the free-particle case. Define two propagators $K_0^{\pm}$ by

$$K_0^+(2; 1) = \theta(t_2 - t_1) \sum_n u_n(2)\bar{u}_n(1),$$

[1] Adapted from Hoyle and Narlikar, 1971. Equation (39) here was (30) there.

$$K_0^-(2; 1) = -\theta(t_1 - t_2) \sum_n u_n(2)\bar{u}_n(1). \tag{39}$$

Both satisfy the inhomogeneous Dirac equation

$$(\nabla\!\!\!/_2 + im)K_0^{\pm}(2; 1) = \delta_4(2, 1), \tag{40}$$

and both are confined within the light cone at point 1, $K_0^+(2; 1)$ being nonzero only for points 2 inside or on the forward half of this cone, and $K_0^-(2; 1)$ being similarly confined to the backward half of the cone.

Corresponding to these propagators we distinguish two types of path, $\Gamma^+{}_{21}$ going forward in time from point 1 to 2 ($t_2 > t_1$) and a path $\Gamma^-{}_{21}$ going backward from 1 to 2 ($t_2 < t_1$). Both are monotonic with respect to time.

To define amplitude along $\Gamma^+{}_{21}$ divide $t_2 - t_1$ into a large number of small subintervals at intermediate points X_i, $i = 0, 1, \ldots, N$. Let $i = 0$ correspond to point 1 and $i = N$ to point 2. Define $P(\Gamma^+{}_{21})$ as

$$\prod_{i=1}^{N} A_i^{-1} K_0^+(i; i - 1), \tag{41}$$

for N large. We shall discuss whether or not N should tend to infinity later. The amplitude $P(\Gamma^-{}_{21})$ for a Γ^- path is similarly defined with a chain of $K_0^-(i; i - 1)$ propagators. Clearly the paths $\Gamma^+{}_{21}$ and $\Gamma^-{}_{21}$ must be timelike throughout.

The measure for the sum over paths is so defined that

$$K_0^{\pm}(2; 1) = \sum_{\Gamma^{\pm}{}_{21}} P(\Gamma^{\pm}{}_{21}) = \int P(\Gamma^{\pm}{}_{21})\, \mathscr{D}^3\Gamma^{\pm}{}_{21}. \tag{42}$$

Suppose we study a free particle in the four-dimensional slab $t_1 \leq t \leq t_2$. At $t = t_1$ paths coming from the past represent an amplitude from some previous history of the particle. We denote this amplitude by ψ_+. Similarly at $t = t_2$ paths coming from the future give an amplitude which we denote by ψ_-. Thus the amplitude at a point 3 within the slab is given by paths $\Gamma^+{}_{31}$ starting at points 1 on $t = t_1$ and weighted by $\psi_+(1)$, together with paths $\Gamma^-{}_{32}$ starting at points 2 on $t = t_2$ and weighted by $\psi_-(2)$. That is to say, the amplitude $\psi(3)$ is given by

$$\psi(3) = \int K_0^+(3; 1)\gamma_4\psi_+(1)\, d^3\boldsymbol{x}_1 - \int K_0^-(3; 2)\gamma_4\psi_-(2)\, d^3\boldsymbol{x}_2. \tag{43}$$

If ψ_+ is made up of positive-energy solutions u_n at $t = t_1$, then ψ_+ for $t > t_1$ is always made up of the same positive-energy solutions. Similarly if ψ_- is made up of negative-energy solutions at $t = t_2$, then ψ_- for $t < t_2$ is always made up of negative-energy solutions. We impose this property as a boundary condition. Then (43) is equivalent to

$$\psi(3) = \int K_+(3; 1)\gamma_4\psi_+(1)\, d^3\boldsymbol{x}_1 - \int K_+(3; 2)\gamma_4\psi_-(2)\, d^3\boldsymbol{x}_2. \qquad (44)$$

Suppose now that we are given an external potential B_i nonzero in $t_1 \leq t \leq t_2$ but zero outside this range. This does not interfere with the boundary condition just discussed. We now admit that within the slab $t_1 \leq t \leq t_2$ interaction with B_i can turn the time-direction of a path. Thus paths within the slab can be zigzag with as many reversals as we like, whereas paths outside the slab are monotonic with respect to t.

The amplitude for a $\Gamma^+{}_{21}$ path is defined by

$$P^B(\Gamma^+{}_{21}) = P(\Gamma^+{}_{21}) \exp\left(-ie\int_{\Gamma^+{}_{21}} B_i\, da^i\right), \qquad (45)$$

and that for a $\Gamma^-{}_{12}$ path (point 1 on $t = t_1$, point 2 on $t = t_2$) by

$$P^B(\Gamma^-{}_{12}) = P(\Gamma^-{}_{12}) \exp\left(-ie\int_{\Gamma^-{}_{12}} B_i\, da^i\right). \qquad (46)$$

Paths within the slab need not be monotonic with respect to t, however. Suppose we have a path from 1 to 2 with $2n$ reversals. Denoting intermediate points by i, sections between successive reversals are monotonic, and the amplitude is given by

$$P^B(\Gamma_{21}) = \prod_i P^B(\Gamma^\pm{}_{i,\,i-1}), \qquad (47)$$

where the $+$ sign holds for forward-going sections and the $-$ sign for the backward-going sections.

Clearly it is also possible to have paths starting at t_1 and ending at t_1. In this case we get a ψ_- at $t = t_1$ even though there may be no ψ_- at $t = t_2$. This does not happen when $B_i = 0$. Similarly we could have a $\psi_+ \neq 0$ at t_2 originating solely from ψ_- at t_2, because of reversals by B_i.

We now wish to calculate the following problem. Given ψ_+ on $t = t_1$ and $\psi_- = 0$ on $t = t_2$, what are ψ_+ on $t = t_2$ and ψ_- on $t = t_1$?

Consider the evaluation of $\psi_+(t_2)$. Formally we have

$$\psi_+{}^B(2) = \int\int P^B(\Gamma_{21})\gamma_4\psi_+(1)\,\mathcal{D}^3\Gamma_{12}\,d^3\boldsymbol{x}_1, \tag{48}$$

which includes paths with reversals. To evaluate (48) consider the paths according to the number of reversals, beginning with paths $\Gamma^+{}_{21}$ without reversals.

Divide the time-range $t_1 \leq t \leq t_2$ into a large number of small intervals. Consider one such interval $(t, t + \epsilon)$. Let the path $\Gamma^+{}_{21}$ intersect the time-sections at t and $t + \epsilon$ at spatial points $\boldsymbol{x}$, $\boldsymbol{y}$, respectively. Since $\Gamma^+{}_{21}$ is timelike, $|\boldsymbol{y} - \boldsymbol{x}| \leq \epsilon$. Expand the exponential factor in (45) in a power series. The unity term gives the usual free-particle contribution. The linear term gives

$$-ie \sum_{\substack{\text{all}\\ \text{intervals}}} \epsilon \int\int [B_4(\boldsymbol{x}, t) - (\boldsymbol{y} - \boldsymbol{x}) \cdot \boldsymbol{B}(\boldsymbol{x}, t)/\epsilon] P(\Gamma^+{}_{21})\gamma_4\psi_+(1)\,\mathcal{D}^3\Gamma^+{}_{21}\,d^3\boldsymbol{x}_1. \tag{49}$$

Now $P(\Gamma^+{}_{21})$ can be written as a product of three factors

$$P(\Gamma^+{}_{21}) = P(\Gamma^+{}_{2y})P(\Gamma^+{}_{yx})P(\Gamma^+{}_{x1}). \tag{50}$$

The sum over $\Gamma^+{}_{21}$ is equivalent to sums over $\Gamma^+{}_{2y}$, $\Gamma^+{}_{yx}$, $\Gamma^+{}_{x1}$, together with integrations over $\boldsymbol{x}$, $\boldsymbol{y}$. The sum over $\Gamma^+{}_{x1}$ propagates the positive-energy part of the wave function at 1 to $\boldsymbol{x}$. This gives $\psi_+(\boldsymbol{x}, t)$. The sum over $P(\Gamma^+{}_{yx})$ gives $K_0{}^+(\boldsymbol{y}, t + \epsilon; \boldsymbol{x}, t)$. The sum over $P(\Gamma^+{}_{2y})$ similarly gives $K_0{}^+(2; \boldsymbol{y}, t + \epsilon)$. The latter function propagates only the forward-going paths at $t + \epsilon$, and the weight of these paths is given by the positive-energy part of the wave function at $(\boldsymbol{y}, t + \epsilon)$. In general, because of B_i in $(t, t + \epsilon)$, there will be a negative-energy part in the wave function at $t + \epsilon$. This negative-energy part gives the weight of backward-going paths "reflected" by the step $(t, t + \epsilon)$. The negative-energy part does not affect $\psi_+(2)$. We ensure that only the positive-energy part of the wave function contributes by using $K_+(2; \boldsymbol{y}, t + \epsilon)$ instead of $K_0{}^+(2; \boldsymbol{y}, t + \epsilon)$. So we get for (49)

$$-ie \sum \epsilon \int\int K_+(2; \boldsymbol{y}, t + \epsilon)\gamma_4[B_4(\boldsymbol{x}, t) - (\boldsymbol{y} - \boldsymbol{x}) \cdot \boldsymbol{B}(\boldsymbol{x}, t)/\epsilon] \cdot K_0{}^+(\boldsymbol{y}, t + \epsilon; \boldsymbol{x}, t)\gamma_4\psi_+(\boldsymbol{x}, t)\,d^3\boldsymbol{x}\,d^3\boldsymbol{y}. \tag{51}$$

Because $|\boldsymbol{y} - \boldsymbol{x}| \leq \epsilon$, we can replace $K_+(2; \boldsymbol{y}, t + \epsilon)$ by $K_+(2; \boldsymbol{x}, t)$ to within a term that tends to zero with ϵ. Using (39) we have

$$K_0{}^+(y, t+\epsilon; x, t) = \sum_n U_n(y)\bar{U}_n(x)e^{-i\epsilon E_n}$$

$$= \delta_3(y, x)\gamma_4 - i\epsilon \sum_n H(y)U_n(y)\bar{U}_n(x) + O(\epsilon^2), \quad (52)$$

where H is the Hamiltonian and

$$u_n(x, t) = U_n(x)e^{-iE_n t}. \quad (53)$$

Thus (51) becomes

$$-ie\sum \epsilon \int K_+(2; x, t)\gamma_4 B_4(x, t)\psi_+(x, t)\, d^3x$$

$$-ie\sum i\epsilon \int\int K_+(2; x, t)\gamma_4(y - x)\cdot B(x, t)$$

$$\cdot\left[\sum_n H(y)U_n(y)\bar{U}_n(x)\right]\gamma_4\psi_+(x, t)\, d^3x\, d^3y. \quad (54)$$

Now for a free particle

$$H \equiv -i\alpha\cdot\nabla + m\gamma_4$$

and the $m\gamma_4$ term makes no contribution to the second part of (54). The only portion of this second part that survives as $\epsilon \to 0$ is that which arises on differentiating $y - x$, using integration by parts with respect to ∇y from $H(y)$. Since $\Sigma_n U_n(y)\bar{U}_n(x) = \delta_3(y, x)\gamma_4$ it is easy to see that (54) is just

$$-ie\sum \epsilon \int K_+(2; x, t)\not{B}(x, t)\psi_+(x, t)\, d^3x. \quad (55)$$

For small ϵ the summation can be replaced by integration with respect to t. Denoting (x, t) as point 3 we therefore obtain

$$-ie\int K_+(2; 3)\not{B}(3)\psi_+(3)\, d\tau_3, \quad (56)$$

the integration with respect to $d\tau_3$ being over the four-dimensional slab $t_1 \leq t \leq t_2$.

The second-order term in the expansion of the exponential in (45) can be worked out in the same way and gives

$$(-ie)^2 \iint_{t_4 > t_3} K_+(2;4)\not{B}(4)K_+(4;3)\not{B}(3)\psi_+(3)\,d\tau_3\,d\tau_4, \tag{57}$$

and similarly for higher-order terms. The appearance of K_+ propagators after each $\not{B}$ can be understood in the following terms. After each $\not{B}$ scattering the wave function appears in a general $\psi_+ + \psi_-$ form. Since we are propagating along a Γ^+ path we must use only the ψ_+ part. That is to say, we need to act on ψ_+. But this is the same as K_+ acting on $\psi_+ + \psi_-$.

Consider now paths with reversals. Evidently to arrive at $t = t_2$ from $t = t_1$ we must have an even number of reversals. It will serve to bring out the main features if we treat the simplest case of two reversals. Consider a path that goes from t_1 to $t_3 > t_1$, then reverses to $t_4 < t_3$, and reverses again to $t_2 > t_4$. In order for paths of this kind to contribute to the amplitude the potential must be nonzero at t_3 and t_4. Again dividing $t_1 \leq t \leq t_2$ into small steps, the integral $\int B_i\, da^i$ is replaced by a sum over steps. This integral appears squared in the second order of the perturbation expansion, so that in this order we have a double sum over steps. The double sum contains a member that refers to the steps $(t_3 - \epsilon, t)$ and $(t_4, t_4 + \epsilon)$. This member gives the lowest-order contribution that the path in question makes to the total amplitude.

We have to sum over all paths from t_1 to t_2 with reversals at t_3 and t_4. Since for such paths the particle is free (in the second order of the perturbation expansion) between t_1 and $t_3 - \epsilon$, it follows that ψ_+ at t_1 is carried into a ψ_+ wave function at $t_3 - \epsilon$. Within the thin slab $(t_3 - \epsilon, t_3)$ scattering by B_i changes the wave function into a $\psi_+ + \psi_-$ combination at t_3. Since the paths are now backward and the particle is again free (second order of the perturbation expansion) between t_3 and $t_4 + \epsilon$, only the ψ_- part of the wave function at t_3 is carried to $t_4 + \epsilon$. Between $t_4 + \epsilon$ and t_4 there is a further scattering by B_i, and this again sets up a $\psi_+ + \psi_-$ combination at t_4. The paths now go forward to t_2, and since the particle is once again free only the ψ_+ part of the wave function at t_4 is carried to t_2. This "accounting" procedure is rather complicated when the $K_0^{\pm}$ propagators are used, but the K_+ propagator automatically keeps a correct record at each reversal. It is therefore preferable to use K_+ for the sections in which the particle is free.

Summing over all paths that reverse at times t_3, t_4 $(t_3 > t_4)$ in $[t_1, t_2]$ we obtain

$$(-ie)^2 \iint_{t_3 > t_4} K_+(2; 4)\not{B}(4)K_+(4; 3)\not{B}(3)\psi_+(3)\, d\tau_3\, d\tau_4, \tag{58}$$

and (57) and (58) together give

$$(-ie)^2 \iint K_+(2; 4)\not{B}(4)K_+(4; 3)\not{B}(3)\psi_+(3)\, d\tau_3\, d\tau_4 \tag{59}$$

as the second-order term of the perturbation expansion.

In this last step we are assuming that the case $t_3 = t_4$ gives no contribution to the amplitude. This case can be dealt with in the following way. When the path first reverses at t_3, the ψ_- part of the wave function is of order ϵ, being proportional to the thickness of the slab $(t_3 - \epsilon, t_3)$. When this ψ_- part is scattered again by the same slab, the resulting ψ_+ part of the wave function is of order ϵ^2. This ψ_+ part then goes to t_2. We have a similar contribution for each thin slab, but since there are only of the order of ϵ^{-1} such slabs, the total "double scattering" is of order ϵ and tends to zero with ϵ.

The above discussion is easily generalized to give

$$\psi_+(2) = \int K_+{}^B(2; 1)\gamma_4\psi_+(1)\, d^3\boldsymbol{x}_1, \tag{60}$$

where

$$K_+{}^B(2; 1) = K_+(2; 1) - ie\int K_+(2; 3)\not{B}(3)K_+(3; 1)\, d\tau_3$$
$$+ (-ie)^2 \iint K_+(2; 4)\not{B}(4)K_+(4; 3)\not{B}(3)K_+(3; 1)\, d\tau_3\, d\tau_4 + \dots . \tag{61}$$

A similar analysis leads to

$$\psi_-(1') = \int K_+{}^B(1'; 1)\gamma_4\psi_+(1)\, d^3\boldsymbol{x}_1, \tag{62}$$

where $K_+{}^B(1'; 1)$ is given by replacing point 2 in (61) by point 1′, the latter point being on $t = t_1$.

The same procedure for (61) as that used for (16) leads easily to

$$[\not{\nabla}_2 + ie\not{B}(2) + im]K_+{}^B(2; 1) = \delta_4(2, 1). \tag{63}$$

Equations (60) and (62) give the solution to the problem we set out to solve, and (61) is the required perturbation expansion.

§A3.2. Many-Particle Interactions and the Generalized Response of the Universe

We begin by restating the conclusions of the preceding section in a slightly different notation and in a somewhat more general form.

We now describe the slab under investigation by $0 \leq t \leq T$ instead of by $t_1 \leq t \leq t_2$. The planes $t = 0$, $t = T$ will be referred to as the surface S and points $1, 2, \ldots, 1', 2', \ldots$ may be anywhere on S. The work of the previous section is described by

$$\psi_{\text{out}}(1') = \int K_+{}^B(1'; 1) \not{n}_1 \psi_{\text{in}}(1)\, dS_1. \tag{64}$$

Here we introduce the concept of ingoing and outgoing parts of the wave function at the surface S. In the notation of the previous section ψ_{in} is a ψ_+ wave function on $t = 0$ and is a ψ_- wave function on $t = T$, while ψ_{out} is a ψ_+ wave function on $t = T$ and a ψ_- wave function on $t = 0$; n^i is the unit inward normal to S, and dS_1 is the surface element at point 1.

Most one-electron problems in quantum electrodynamics are concerned with calculating ψ_{out} when ψ_{in} and the external field B_i are given.

Consider the following generalization of (64) to the many-particle problem

$$\Psi_{\text{out}}(1', 2, \ldots) = \int\int \ldots \int K_+{}^B(1'; 1) \not{n}_1 K_+{}^B(2'; 2) \not{n}_2 \ldots \Psi_{\text{in}}(1, 2, \ldots)\, dS_1\, dS_2 \ldots. \tag{65}$$

The wave function Ψ_{in} now has four components for each particle and has a point $1, 2, \ldots$ for each particle. Similarly Ψ_{out} has four components and a member of $1', 2', \ldots$ for each particle. The matrix $K_+{}^B(1'; 1)\not{n}_1$ acts on the components of Ψ_{in} for the particle at point 1; $K_+{}^B(2'; 2)\not{n}_2$ acts on the components for the particle at point 2; and so on. The order in which these matrices are written is irrelevant.

This generalization is valid for particles acted on by an external field, but is invalid when there are mutual interactions between the particles. Then we write

$$\Psi_{\text{out}}(1', 2', \ldots) = \int\int \ldots \int K(1', 2', \ldots; 1, 2, \ldots) \not{n}_1 \not{n}_2 \ldots \Psi_{\text{in}}(1, 2, \ldots)\, dS_1\, dS_2 \ldots \tag{66}$$

in which the many-particle propagator $K(1', 2', \ldots; 1, 2, \ldots)$ is determined in the following way.

We recall that

$$K_+{}^B(1'; 1) = \int P(\Gamma_{1'1}) \exp\left(-ie_a \int_{\Gamma_{1'1}} B_i\, da^i\right) \mathscr{D}^3\Gamma_{1'1}. \qquad (67)$$

The multiple-path integral for a many-particle system in an external field B_i is

$$K(1', 2', \ldots; 1, 2, \ldots)$$

$$= \int \cdots \int P(\Gamma_{1'1})P(\Gamma_{2'2}) \ldots \exp\left(-ie_a \int_{\Gamma_{1'1}} B_i\, da^i - ie_b \int_{\Gamma_{2'2}} B_i\, db^i - \ldots\right)$$
$$\times \mathscr{D}^3\Gamma_{1'1} \mathscr{D}^3\Gamma_{2'2} \cdots \qquad (68)$$

and this separates into the product $K_+{}^B(1'; 1)\, K_+{}^B(2; 2) \ldots$ as stated above. But when the particles are mutually interacting (68) is changed to

$$K(1', 2', \ldots; 1, 2, \ldots) = \int\int \cdots \int P(\Gamma_{1'1})P(\Gamma_{2'2}) \ldots \exp(iR)$$
$$\cdot \exp\left[-i\left(e_a \int_{\Gamma_{1'1}} B_i\, da^i + \ldots\right)\right] \mathscr{D}^3\Gamma_{1'1} \mathscr{D}^3\Gamma_{2'2} \cdots, \qquad (69)$$

in which the factor $\exp(iR)$ represents interparticle action. In general $\exp(iR)$ takes such a form as to prevent (69) from being separable. The particles are here denoted by $a, b, \ldots,$ and $da_i, db_i, \ldots$ are coordinate displacements along the paths $\Gamma_{1'1}, \Gamma_{2'2}, \ldots,$ respectively.

We now aim to prove that all the results of the usual theory of quantum electrodynamics follow from choosing the classical time-symmetric action for R,

$$R = -\frac{1}{2} \sum_a \sum_b e_a e_b \int_{\Gamma_{1'1}} \int_{\Gamma_{2'2}} \delta(q^2{}_{AB})\, da_i\, db^i. \qquad (70)$$

Before proceeding, however, we pause to note that if $\Psi_{\text{in}}\,(1, 2, \ldots)$ is antisymmetric with respect to every pair of particles, then so is $\Psi_{\text{out}}\,(1', 2', \ldots)$. This can be seen from (66), (69), and (70). The exclusion principle has a rather mysterious role in the usual formulation of quantum electrodynamics. The present path-integral approach throws some light on its meaning. It prevents the paths of different particles from sharing common points. The paths $\Gamma_{1'1}, \Gamma_{2'2},$ cannot cross, for example—more accurately, paths with common points make zero total contribution to the amplitude.

Paths for the same particle, on the other hand, can have common points or common segments, because paths for the same particle are alternatives—the particle follows one of its possible paths. It is indeed this property that permits us to say whether two paths belong to the same particle or to different particles.

Returning to the multiple-path integral (69), the electromagnetic factor is the exponential of i times

$$-\sum_a e_a \int B_i\, da^i - \frac{1}{2}\sum_a \sum_b e_a e_b \int\int \delta(q^2{}_{AB})\, da^i\, db_i, \tag{71}$$

the line integrals being over the paths $\Gamma_{1'1}$, $\Gamma_{2'2}, \ldots,$ for the particles a, $b, \ldots,$ respectively. The paths are segments lying within the slab $0 \leq t \leq T$. There are interactions with paths outside the slab, and it is these interactions that give B_i. There are contributions to B_i from both $t < 0$ and $t > T$. The interaction with the future arises because paths interact through the time-symmetric function $\delta(q^2)$. The external field from $t < 0$ has already been discussed in detail in §4.3, so we shall not repeat the discussion here. That is to say

$$B^i = B^i{}_{t<0} + B^i{}_{t>T}, \tag{72}$$

and we take

$$B^i{}_{t<0} = 0. \tag{73}$$

The first term of (71) is then the response of the universe.

Apart from containing terms with $a = b$, which we shall discuss at the end of the present section, (71) is the same as the classical action in the direct-particle theory. It is interesting to recall the classical treatment of the action, since the classical treatment and the quantum treatment will turn out to have similar features. First we write

$$-\frac{1}{2}\sum_{a \neq b}\sum e_a e_b \int\int \delta(q^2{}_{AB})\, da^i\, db_i = -\frac{1}{2}\sum_a e_a \int A^i{}_{(a)}\, da_i,$$

where

$$A^i{}_{(a)}(X) = \sum_{b \neq a} e_b \int \delta(q^2{}_{XB})\, db^i. \tag{74}$$

Separating $\delta(q^2{}_{AB})$ into advanced and retarded parts,

$$\delta(q^2{}_{AB}) = \frac{1}{2|\boldsymbol{x}_A - \boldsymbol{x}_B|}[\delta(t_A - t_B - |\boldsymbol{x}_A - \boldsymbol{x}_B|) + \delta(t_A - t_B + |\boldsymbol{x}_A - \boldsymbol{x}_B|)], \quad (75)$$

and writing

$$A^i{}_{(a)}(X) = \frac{1}{2}[A^i{}_{(a)}(X)_{\text{ret}} + A^i{}_{(a)}(X)_{\text{adv}}], \quad (76)$$

$$A^i{}_{(a)}(X)_{\text{ret}} = \sum_{b \neq a} A^{i(b)}(X)_{\text{ret}} = \sum_{b \neq a} e_b \int \frac{\delta(t_X - t_B - |\boldsymbol{x} - \boldsymbol{x}_B|)}{|\boldsymbol{x} - \boldsymbol{x}_B|} db^i, \quad (77)$$

$$A^i{}_{(a)}(X)_{\text{adv}} = \sum_{b \neq a} A^{i(b)}(X)_{\text{adv}} = \sum_{b \neq a} e_b \int \frac{\delta(t_X - t_B + |\boldsymbol{x} - \boldsymbol{x}_B|)}{|\boldsymbol{x} - \boldsymbol{x}_B|} db^i, \quad (78)$$

enables the classical action to be written in the form

$$-\sum_a e_a \int B^i{}_{t>T}\, da_i - \frac{1}{2}\sum_a e_a \int [A^i{}_{(a)\text{ret}} + A^i{}_{(a)\text{adv}}]\, da_i. \quad (79)$$

The classical response condition is

$$B_i(X)_{t>T} = \frac{1}{2}\sum_a [A_i{}^{(a)}(X)_{\text{ret}} - A_i{}^{(a)}(X)_{\text{adv}}]. \quad (80)$$

Eliminating $B^i{}_{t>T}$ between (79) and (80), we get

$$-\sum_a e_a \int \left\{A^i{}_{(a)\text{ret}} + \frac{1}{2}[A^{i(a)}{}_{\text{ret}} - A^{i(a)}{}_{\text{adv}}]\right\} da_i. \quad (81)$$

The first term in the curly brackets is the usual retarded classical field of all particles other than a, and the second term is the usual radiation reaction on a.

Our aim is to obtain a generalized response condition for the quantum case, analogous to (80), that together with (71) leads to an electromagnetic phase factor in (69) giving all the results of the usual theory of quantum electrodynamics.

The quantum case differs from classical theory in that conjugate paths,

$\boldsymbol{a}'(t)$ for particle a, $\boldsymbol{b}'(t)$ for particle $b, \ldots$, must be introduced. These conjugate paths go from points $1^*, 2^*, \ldots$, on S to the same points $1', 2', \ldots$, as do the paths $\boldsymbol{a}(t)$, $\boldsymbol{b}(t) \ldots$. Instead of (71) we now have (71) minus the contribution of the conjugate paths. It is easy to see that this combination of paths and conjugate paths gives

$$-\sum_a e_a \left\{ \int \left[\frac{1}{2} (A^i{}_{\text{ret}} + A^i{}_{\text{adv}}) + B^i{}_{t>T} \right] da_i \right.$$
$$\left. - \int \left[\frac{1}{2} (A'^i{}_{\text{ret}} + A'^i{}_{\text{adv}}) + B'^i{}_{t>T} \right] da_i' \right\} \quad (82)$$

in which

$$A^i(X)_{\text{ret}} = \sum_b e_b \int \frac{\delta(t_X - t_B - |\boldsymbol{x} - \boldsymbol{x}_B|)}{|\boldsymbol{x} - \boldsymbol{x}_B|} db^i, \quad (83)$$

$$A^i(X)_{\text{adv}} = \sum_b e_b \int \frac{\delta(t_X - t_B + |\boldsymbol{x} - \boldsymbol{x}_B|)}{|\boldsymbol{x} - \boldsymbol{x}_B|} db^i, \quad (84)$$

and where the primed fields refer to the conjugate paths and are given by expressions similar to (83), (84), and $B'^i{}_{t>T}$ is the response to the conjugate paths. Multiplying (82) by i and taking the exponential gives the electromagnetic phase factor.

The first integral in (82) cannot be expressed in a form suited to practical calculation except in association with the second integral. That is to say, we do not know the response of the universe except in a probability calculation. The usual theory of quantum electrodynamics, on the other hand, allows a complete calculation of the amplitude alone, irrespective of whether a probability is taken. Since amplitudes do not appear in physical statements, the usual theory seems unsatisfactory in this respect. It is true that physical statements can be made by discarding the phase of the amplitude, but then why go to the trouble to calculate the phase if it is not required?

The slab $0 \leq t \leq T$ now has an explicit meaning. It is the slab in which an experiment is performed, i.e., for which we require a physical statement.

Paths and conjugate paths together permit a separation of positive and negative frequencies. In classical theory there is one path for each particle, and this one path contributes both positive and negative frequencies. In quantum electrodynamics paths give positive frequencies and conjugate paths give negative frequencies. Negative and positive frequencies are introduced in the following way. Define

$$A^i(X)_{\text{ret}\pm} = \sum_b e_b \int \frac{\delta_\pm(t_X - t_B - |\boldsymbol{x} - \boldsymbol{x}_B|)}{2|\boldsymbol{x} - \boldsymbol{x}_B|} \, db^i, \tag{85}$$

$$A^i(X)_{\text{adv}\pm} = \sum_b e_b \int \frac{\delta_\pm(t_X - t_B + |\boldsymbol{x} - \boldsymbol{x}_B|)}{2|\boldsymbol{x} - \boldsymbol{x}_B|} \, db^i, \tag{86}$$

where

$$\delta_\pm(x) = \frac{1}{\pi} \int_0^\infty e^{\mp i\omega x} \, d\omega = \delta(x) \mp \frac{i}{\pi} \frac{\mathcal{P}}{x}. \tag{87}$$

It may be noted that whereas the sum of $A^i{}_{\text{ret}+}$, $A^i{}_{\text{ret}-}$, is a vector, namely, the classical retarded field

$$A^i(X)_{\text{ret}} = A^i{}_{\text{ret}+} + A^i{}_{\text{ret}-}, \tag{88}$$

$A^i{}_{\text{ret}+}$, $A^i{}_{\text{ret}-}$ are not vectors by themselves. Similarly

$$A^i(X)_{\text{adv}} = A^i{}_{\text{adv}+} + A^i{}_{\text{adv}-} \tag{89}$$

is the classical advanced field, but $A^i{}_{\text{adv}+}$, $A^i{}_{\text{adv}-}$, are not vectors by themselves.

This separation into positive and negative frequencies permits two new vectors to be defined, $A^i{}_{\text{ret}+} - A^i{}_{\text{adv}+}$ and $A^i{}_{\text{ret}-} - A^i{}_{\text{adv}-}$. These vectors appear in quantum electrodynamics but not in classical theory. It is straightforward to show that

$$A^i(X)_{\text{ret}+} - A^i(X)_{\text{adv}+}$$

$$= \sum_b e_b \left[\int_{t_X > t_B} \delta_+(q^2{}_{XB}) \, db^i - \int_{t_B > t_X} \delta_-(q^2{}_{BX}) \, db^i \right], \tag{90}$$

$$A^i(X)_{\text{ret}-} - A^i(X)_{\text{adv}-}$$

$$= \sum_b e_b \left[\int_{t_X > t_B} \delta_-(q^2{}_{XB}) \, db^i - \int_{t_B > t_X} \delta_+(q^2{}_{BX}) \, db^i \right]. \tag{91}$$

Although $\delta_\pm(q^2)$ are invariant functions, it is not immediately clear that (90) and (91) are vectors, because of the time restrictions. However, the contributions of the principal parts in $\delta_\pm(q^2)$ have opposite signs. It is easy to see therefore that the time restrictions do not apply to the principal parts, which accordingly give vectors. The time restrictions continue to apply to the $\delta(q^2)$ part of $\delta_\pm(q^2)$, but since $\delta(q^2)$ is confined to the light cone, the restrictions are invariant, and the $\delta(q^2)$ part also gives vectors.

The quantum response corresponding to (80) is

$$B_i(X)_{t>T}$$

$$= \frac{1}{2} \sum_b \{[A_i^{(b)}(X)_{\text{ret}+} - A_i^{(b)}(X)_{\text{adv}+}] + [A_i'^{(b)}(X)_{\text{ret}-} - A_i'^{(b)}(X)_{\text{adv}-}]\}. \quad (92)$$

As stated above, the paths carry the positive frequencies and the conjugate paths carry the negative frequencies. If we imagine a coalescence of paths and conjugate paths (92) becomes the same as (80). To obtain an equivalent condition for $B_i'(X)_{t>T}$ we interchange paths with conjugate paths and positive frequencies with negative frequencies. This double interchange leaves (92) unaltered, so that

$$B_i'(X)_{t>T} = B_i(X)_{t>T}. \quad (93)$$

To complete the work of the present section we calculate the contribution to (82) from particles a, b. For $a \neq b$ we have four paths $\boldsymbol{a}$, $\boldsymbol{b}$, $\boldsymbol{a}'$, $\boldsymbol{b}'$, which combine in the pairs $(\boldsymbol{a}, \boldsymbol{b})$, $(\boldsymbol{a}', \boldsymbol{b}')$, $(\boldsymbol{a}', \boldsymbol{b})$, $(\boldsymbol{a}, \boldsymbol{b}')$. The contribution of $(\boldsymbol{a}, \boldsymbol{b})$ has two terms,

$$-e_a e_b \int\int \delta(q^2{}_{AB})\, da_i\, db^i, \quad (94)$$

from the time symmetric interparticle action and

$$-\frac{1}{2} e_a e_b \left\{ \int\int_{t_A > t_B} [\delta_+(q^2{}_{AB}) - \delta_-(q^2{}_{AB})]\, da_i\, db^i \right.$$

$$\left. + \int\int_{t_B > t_A} [\delta_+(q^2{}_{BA}) - \delta_-(q^2{}_{BA})]\, da_i\, db^i \right\} \quad (95)$$

from B_i. Since

$$2\delta(q^2{}_{AB}) = \delta_+(q^2{}_{AB}) + \delta_-(q^2{}_{AB}), \qquad \delta_\pm(q^2{}_{AB}) = \delta_\pm(q^2{}_{BA}), \quad (96)$$

the sum of (94) and (95) is

$$-e_a e_b \int\int \delta_+(q^2{}_{AB})\, da_i\, db^i. \quad (97)$$

The pair $(\boldsymbol{a}', \boldsymbol{b}')$ similarly give

$$+e_a e_b \int\int \delta_-(q^2_{A'B'})\, da_i{}'\, db'^i. \tag{98}$$

Only the B_i, $B_i{}'$ terms in (82) contribute to $(\boldsymbol{a}', \boldsymbol{b})$. It is not hard to show that these terms combine to give

$$e_a e_b \left[\int\int_{t_{A'} > t_B} \delta_+(q^2{}_{A'B})\, da_i{}'\, db^i - \int\int_{t_B > t_{A'}} \delta_-(q^2{}_{BA'})\, da_i{}'\, db^i \right]. \tag{99}$$

The pair $(\boldsymbol{a}, \boldsymbol{b}')$ gives

$$e_a e_b \left[\int\int_{t_{B'} > t_A} \delta_+(q^2{}_{B'A})\, da^i\, db_i{}' - \int\int_{t_A > t_{B'}} \delta_-(q^2{}_{AB'})\, da^i\, db_i{}' \right]. \tag{100}$$

The total contribution of the particles a, b, to the electromagnetic phase factor is given by summing (97)–(100), multiplying by i, and taking the exponential,

$$\begin{aligned} \exp\Bigg\{ ie_a e_b \Bigg[&\int\int_{t_{A'} > t_B} \delta_+(q^2{}_{A'B})\, da_i{}'\, db^i - \int\int_{t_B > t_{A'}} \delta_-(q^2{}_{BA'})\, da_i{}'\, db^i \\ &+ \int\int_{t_{B'} > t_A} \delta_+(q^2{}_{B'A})\, da^i\, db_i{}' - \int\int_{t_A > t_{B'}} \delta_-(q^2{}_{AB'})\, da^i\, db_i{}' \\ &- \int\int \delta_+(q^2{}_{AB})\, da^i\, db_i + \int\int \delta_-(q^2{}_{A'B'})\, da_i{}'\, db'^i \Bigg] \Bigg\}. \end{aligned} \tag{101}$$

This is the influence functional $F[\boldsymbol{a}, \boldsymbol{b}; \boldsymbol{a}', \boldsymbol{b}']$ contributed by the particles a, b. It can be expressed in terms of three-dimensional Fourier integrals using

$$\delta_\pm(q^2{}_{21}) = \pm 4\pi i \int \frac{1}{2K} \exp[\mp iK|t_2 - t_1| \pm i\boldsymbol{k}\cdot(\boldsymbol{x}_2 - \boldsymbol{x}_1)] \frac{d^3\boldsymbol{k}}{(2\pi)^3}, \tag{102}$$

where $K = |\boldsymbol{k}|$ and $q^2{}_{21}$ is the invariant distance between points 1 and 2. We get

$F[\boldsymbol{a}, \boldsymbol{b}; \boldsymbol{a}', \boldsymbol{b}']$

$$= \exp\left(\frac{e_a e_b}{4\pi^2} \int d\Omega \int K\, dK \left\{ \int\int \exp[-iK|t_A - t_B| + i\boldsymbol{k}\cdot(\boldsymbol{x}_B - \boldsymbol{x}_A)]\, da_i\, db^i \right.\right.$$

$$+\int\int \exp[iK|t_{A'} - t_{B'}| + i\boldsymbol{k}\cdot(\boldsymbol{x}_{A'} - \boldsymbol{x}_{B'})]\, da_i'\, db'^i$$

$$-\int\int \exp[iK(t_A - t_{B'}) + i\boldsymbol{k}\cdot(\boldsymbol{x}_{B'} - \boldsymbol{x}_A)]\, da_i\, db'^i$$

$$-\int\int \exp[iK(t_B - t_{A'}) + i\boldsymbol{k}\cdot(\boldsymbol{x}_{A'} - \boldsymbol{x}_B)]\, da_i'\, db^i\Big\}\Big). \tag{103}$$

The different terms in (103) can be interpreted as follows. For $t_A > t_B$ the positive terms in the curly brackets contribute to a downward transition of particle b and an upward transition of particle a, and *vice versa* for $t_B > t_A$, whereas the negative terms contribute to downward transitions of both a and b. Thus (103) refers to absorption and stimulated emission. Since the coefficient in front of all terms in the exponential is the same, the probabilities for these two processes are the same. The influence functional (103) leads to all the usual results for the interaction of two charged particles. There is a similar factor in the multiple-path integral (69) for every pair of particles.

Consider next the case $a = b$. There are contributions from three path combinations $(\boldsymbol{a}, \boldsymbol{a}')$, $(\boldsymbol{a}, \boldsymbol{a})$, and $(\boldsymbol{a}', \boldsymbol{a}')$. These may be obtained in the same way as in the above discussion, or by identifying $\boldsymbol{a}$ and $\boldsymbol{b}$, $\boldsymbol{a}'$ and $\boldsymbol{b}'$, in the above formulas. However, if we follow the latter course a factor $\frac{1}{2}$ must be introduced, because of the $\frac{1}{2}$ in (70) and because $\boldsymbol{a}$ and $\boldsymbol{a}'$ interact only once. The result is

$$F[\boldsymbol{a}, \boldsymbol{a}'] = \exp\Big\{ie_a^2\Big[\int\int_{t_{A'} > t_A} \delta_+(q^2_{A'A})\, da_i'\, da^i$$
$$-\int\int_{t_A > t_{A'}} \delta_-(q^2_{AA'})\, da_i'\, da^i$$
$$-\frac{1}{2}\int\int \delta_+(q^2_{A\tilde{A}})\, da^i\, d\tilde{a}_i + \frac{1}{2}\int\int \delta_-(q^2_{A'\tilde{A}'})\, da'^i\, d\tilde{a}_i'\Big]\Big\}. \tag{104}$$

In terms of Fourier integrals the corresponding result is

$$F[\boldsymbol{a}, \boldsymbol{a}'] = \exp\Big[\frac{e_a^2}{4\pi^2}\int d\Omega \int_0^\infty k\, dK\Big\{-\int\int \exp[iK(t_A - t_{A'})$$
$$+ i\boldsymbol{k}\cdot(\boldsymbol{x}_{A'} - \boldsymbol{x}_A)]\, da^i\, da_i'$$
$$+\int\int_{t_A > t_{\tilde{A}}} \exp[iK(t_{\tilde{A}} - t_A) + i\boldsymbol{k}\cdot(\boldsymbol{x}_{\tilde{A}} - \boldsymbol{x}_A)]\, da^i\, d\tilde{a}_i$$
$$+\int\int_{t_{A'} > t_{\tilde{A}'}} \exp[iK(t_{A'} - t_{\tilde{A}'}) + i\boldsymbol{k}\cdot(\boldsymbol{x}_{\tilde{A}'} - \boldsymbol{x}_{A'})]\, da'^i\, d\tilde{a}_i'\Big\}\Big]. \tag{105}$$

The first term in the curly bracket gives spontaneous transitions, whereas the second and third terms contribute "radiative correction effects." There is a factor of the form (105) in the multiple-path integral (69) for every particle.

For many purposes it is preferable to express the $\delta_{\pm}(q^2)$ functions in (101) and (104) as four-dimensional Fourier integrals, using

$$\delta_{\pm}(q^2) = -4\pi \int_{\epsilon \to 0+} \frac{1}{k^2 \pm i\epsilon} e^{-ik\cdot(x_2 - x_1)} \frac{d^4k}{(2\pi)^4}. \tag{106}$$

We come now to the problem of the self-action contributions to (71). The contribution of particle a is

$$-\frac{1}{2} e_a{}^2 \int\int \delta(q^2{}_{A\tilde{A}})\, da^i\, d\tilde{a}_i, \tag{107}$$

where A, $\tilde{A}$, both lie on the path $\boldsymbol{a}(t)$. In classical theory (107) is omitted entirely; particles do not act on themselves. It is actually sufficient in classical theory to make a weaker hypothesis—that sufficiently small steps of a path do not act on themselves—for classical paths are wholly timelike, and $q^2{}_{A\tilde{A}} > 0$, giving $\delta(q^2{}_{A\tilde{A}}) = 0$ when A and $\tilde{A}$ are separated.

In quantum theory, on the other hand, there is a difference between omitting (107) and making the hypothesis that individual path steps do not act on themselves, since the latter hypothesis still permits different Γ^+ and Γ^- sections of a path to interact, whereas a complete omission of (107) would remove all self-interaction. In the above work we include (107) in order to obtain the usual electron–positron interaction. If there were no self-action, the response $B^i{}_{t>T}$ would contribute the exponential of

$$-\frac{1}{2} ie_a \int [A^{(a)}{}_{i\text{ret}+} - A^{(a)}{}_{i\text{adv}+}]\, da^i$$

$$= -ie_a \int A^{(a)}{}_{i\text{ret}+}\, da^i + \frac{1}{2} ie_a \int [A^{(a)}{}_{i\text{ret}+} + A^{(a)}{}_{i\text{adv}+}]\, da^i \tag{108}$$

to the $(\boldsymbol{a}, \boldsymbol{a})$ term in (105). The second term in (108) is minus the self-action contribution (107). It is canceled therefore when (107) is included, so that the first term of (108) yields the $(\boldsymbol{a}, \boldsymbol{a})$ part of (105). If (107) is taken to be zero for timelike paths, the second term of (108) is also zero, and we still obtain (105), even though self-action is absent.

It seems possible that in quantum theory (107) may vanish for Γ^+ or Γ^- paths individually, even without requiring the hypothesis that individual path steps do not act on themselves. Suppose A and $\tilde{A}$ are on the same step, $da^i \equiv d\tilde{a}^i$. Since quantum paths are made up of null segments, $q^2{}_{A\tilde{A}} = 0$, but this does not cause difficulty so long as A and $\tilde{A}$ are distinct, since $da_i\, d\tilde{a}^i$ is proportional to $q^2{}_{A\tilde{A}}$ and $q^2{}_{A\tilde{A}}\, \delta(q^2{}_{A\tilde{A}}) = 0$. In general for an individual Γ^+ or Γ^- path, either we have $q^2{}_{A\tilde{A}} > 0$, $\delta(q^2{}_{A\tilde{A}}) = 0$, or the path between A and $\tilde{A}$ is a straight null line. So long as A and $\tilde{A}$ are taken to be interior prints within da^i, $d\tilde{a}^i$, we have $da_i\, d\tilde{a}^i$ proportional to $q^2{}_{A\tilde{A}}$ and $\delta(q^2{}_{A\tilde{A}})\, da_i\, d\tilde{a}^i$ is again zero.

The self-action integral becomes ambiguous, however, when A and $\tilde{A}$ are permitted to coincide. This ambiguity is removed if we omit self-action on a sufficiently small scale. We shall find in §A3.4 that this question is intimately connected with divergences in radiation correction processes.

§A3.3. Vacuum Loops

The amplitude for any process in quantum electrodynamics can be calculated from a perturbation expansion, provided a suitable factor, say, C_v, is included to represent vacuum loops. In §A3.1 we obtained the perturbation expansion for a single particle in an external field by the path-integral method. In §A3.2 this method was extended to a many-particle system, and the mutual interactions of the particles were correctly represented by combining the classical electromagnetic action with the response of the universe. This is the position we have reached. To complete our program it remains to return to the problem of vacuum loops and to work out the factor C_v.

We show first that C_v can be obtained by applying an antisymmetrization process to the perturbation expansion (61) for a particle in an external field. Since (61) was obtained from a path integral, it is necessary to show next how the extra terms required by antisymmetrization can also be represented by a path integral. This one-particle treatment must then be generalized to many particles and must include interparticle action. We proceed in these three stages.

The idea underlying the antisymmetrization of (61) is illustrated in general terms in Figure 4.5, p. 76. In (a) we have two particles going forward in time, one particle going from 1 to 3 by a Γ^+ path, the other from 4 to 2 also by a Γ^+ path. The amplitude for (a) does not represent the total amplitude for there to be a particle at 2 and a particle at 3. The exclusion principle requires subtraction of (b), with 1 going to 2 and 4 to 3.

In (c) we have a single-particle path with two time reversals, and with an external field B_i acting at the points 3 and 4 of reversal. The configuration of points is the same in (c) as in (a), and the electron paths from 1 to 3,

4 to 2 are the same. If now we follow the suggestion of the preceding section—that time reversals generate "new" particles—we require the Γ^+ sections from 1 to 3, 4 to 2 in (c) to be on the same footing as the same two paths in (a). We should then subtract the amplitude for the possibility that the Γ^+ paths starting from 1 and 4 go to 2 and 3, as they do in (b), instead of to 3 and 2. The effect of this redirection of the paths is shown in (d). We now have a single Γ^+ path 1 to 2 and a disconnected closed loop with a Γ^+ path from 4 to 3 and a Γ^- path from 3 to 4. The external field acts at the two points of reversal of the loop, not on the Γ^+ path from 1 to 2. We shall next give quantitative expression to this idea.

The amplitude for (c) is given by restricting the second-order term in the perturbation expansion to $t_3 > t_4$:

$$(-ie)^2 \iint_{t_3 > t_4} K_+(2;4)\not{B}(4)K_+(4;3)\not{B}(3)K_+(3;1)\,d\tau_3\,d\tau_4. \tag{109}$$

Since 1 to 3 is a Γ^+ path, (109) operates on a ψ_+ wave function at point 1. Writing the wave function as $v_+(1)$ and integrating spatially with respect to 1, the amplitude for the double scattering of (c) is

$$(-ie)^2 \iint_{t_3 > t_4} K_+(2;4)\not{B}(4)K_+(4;3)\not{B}(3)v_+(3)\,d\tau_3\,d\tau_4. \tag{110}$$

The spatial integration with respect to 1 is over the $t = 0$ face of the slab $0 = t_1 \leq t \leq t_2 = T$, the integrations with respect to $d\tau_3$, $d\tau_4$, are taken through the slab, and point 2 is on $t = T$. Since $t_2 > t_4$ we have

$$K_+(2;4) = \sum_{E_n > 0} u_n(2)\bar{u}_n(4), \tag{111}$$

and we can write (110) in the form

$$(-ie)^2 \sum_{E_n > 0} u_n(2) \iint_{t_3 > t_4} \bar{u}_n(4)\not{B}(4)K_+(4;3)\not{B}(3)v_+(3)\,d\tau_3\,d\tau_4. \tag{112}$$

Next we have to subtract the amplitude for (d). This is

$$-(-ie)^2 \sum_{E_n > 0} v_+(2) \iint_{t_3 > t_4} \bar{u}_n(4)\not{B}(4)K_+(4;3)\not{B}(3)u_n(3)\,d\tau_3\,d\tau_4. \tag{113}$$

The sum of (112) and (113) is antisymmetric with respect to interchange of u_n and v_+, as was required by the qualitative discussion based on Figure 4.5.

Now we use the matrix relation

$$\sum_{\text{spins}} \bar{u}_n(4)\not{\beta}(4)K_+(4;3)\not{\beta}(3)u_n(3)$$
$$= \text{Tr}\left[\sum_{\text{spins}} u_n(3)\bar{u}_n(4)\not{\beta}(4)K_+(4;3)\not{\beta}(3)\right] \quad (114)$$

together with

$$K_+(3;4) = \sum_{E_n>0} u_n(3)\bar{u}_n(4), \qquad t_3 > t_4 \quad (115)$$

and obtain for (113),

$$-(-ie)^2 v_+(2) \iint_{t_3>t_4} \text{Tr}[K_+(3;4)\not{\beta}(4)K_+(4;3)\not{\beta}(3)]\, d\tau_3\, d\tau_4. \quad (116)$$

The restriction $t_3 > t_4$ can be removed provided $\frac{1}{2}$ is introduced, and the resulting coefficient of $v_+(2)$ is

$$-\frac{1}{2}(-ie)^2 \iint \text{Tr}[K_+(3;4)\not{\beta}(4)K_+(4;3)\not{\beta}(3)]\, d\tau_3\, d\tau_4. \quad (117)$$

Apart from the unity term, this is the lowest-order term in C_v. Asymmetrization of higher-order terms in the perturbation expansion proceeds in the same way. In third order there are new terms in which the potential scatters v_+ and also acts twice in a closed loop. In fourth order the new terms scatter v_+ twice and act twice in a closed loop, act four times in a closed loop, and act twice in each of two loops. The net effect of all terms of the perturbation expansion is to give $C_v = \exp(-L)$, say.

Next we see how (117) and other terms arising from the perturbation expansion can be represented by path integrals. The loop Γ^0 is made up of $\Gamma^{\pm}$ sections, as was the path Γ_{21} in (47). The amplitude for a loop can be obtained from (47) by identifying points 1 and 2, and taking the trace of the righthand side,

$$P^B(\Gamma^0) = \text{Tr}\left[\prod_i P^B(\Gamma^{\pm}{}_{i,i-1})\right]. \quad (118)$$

By summing loops in the slab $0 \le t \le T$ we obtain the path integral

$$\int P^B(\Gamma^0)\,\mathscr{D}^3\Gamma^0. \tag{119}$$

The field B_i is a simple exponential factor in each $P^B(\Gamma^{\pm}{}_{i,i-1})$ factor on the righthand side of (118). Taken together these factors give the integral of B_i round Γ^0. We write

$$P^B(\Gamma^0) = \exp\left[-ie\oint_{\Gamma^0} B_i\,dl^i\right] P(\Gamma^0), \tag{120}$$

where

$$P(\Gamma^0) = \mathrm{Tr}\left[\prod_i P(\Gamma^{\pm}{}_{i,i-1})\right]. \tag{121}$$

It is remarkable that $-L$ in $\exp(-L) = C_v$ is just minus the simple path integral (119),

$$-L = -\int P(\Gamma^0)\exp\left[-ie\oint_{\Gamma^0} B_i\,dl^i\right]\mathscr{D}^3\Gamma^0, \tag{122}$$

where dl^i is a coordinate displacement along Γ^0.

Equation (122) is the path integral, with the correct sign, for one loop. The path integral for n closed loops, $l, m, \ldots$, is

$$\frac{1}{n!}(-1)^n \int \cdots \int P(\Gamma_l{}^0)P(\Gamma_m{}^0)$$

$$\ldots \exp\left[-ie\oint_{\Gamma_l{}^0} B_i\,dl^i - ie\oint_{\Gamma_m{}^0} B_i\,dm^i - \ldots\right]\mathscr{D}^3\Gamma_l{}^0\,\mathscr{D}^3\Gamma_m{}^0 \ldots. \tag{123}$$

This may be compared with (68) for the particles $a, b, \ldots$, with the open paths $\Gamma_{1'1}, \Gamma_{2'2}, \ldots$. The factor $1/n!$ arises because the set of loops l, $m, \ldots$, is counted $n!$ times in the summations.

Before continuing with (123) we pause to note that the amplitudes $P(\Gamma_l{}^0)$ are determined in terms of the $\Gamma^{\pm}{}_{i,i-1}$ paths which were defined in §A3.1 in terms of the $K_0{}^{\pm}$ propagators, whereas it is usual to involve the K_+ propagators. The use of the K_+ propagators is due to the trace property of $P(\Gamma^0)$. Thus the trace can be represented by a summation with respect to wave functions,

$$P(\Gamma^0) = \sum \bar{u}\left[\prod P(\Gamma^{\pm}{}_{i,i-1})\right]u, \tag{124}$$

where the summation is over normalized states for any momentum $\boldsymbol{p}$. We require both $E = \pm\sqrt{\boldsymbol{p}^2 + m^2}$. The spacetime dependence of u cancels that of $\bar{u}$ because the wave functions are taken at the "beginning" and "ending" of the loop, i.e., at the same point. Remembering that the Γ^+ sections of Γ^0 must be weighted by a ψ_+ wave function, and the Γ^- sections by a ψ_- wave function, the K_+ propagator is needed to keep correct accounting, exactly as in the derivation of (61).

We have to generalize our path integral so that it includes both particles $a, b, \ldots$, and loops $l, m, \ldots$, and also so that mutual interactions are present as well as an external field. The path integral is to give the many-particle propagator $K(1', 1; 2', 2; \ldots)$. The expression (69) for K applies when there are no loops. When there are n loops $l, m, \ldots$, there is a contribution to K given by

$$\begin{aligned}
&\frac{1}{n!}(-1)^n \int \ldots \int P(\Gamma_{1'1})P(\Gamma_{2'2}) \ldots P(\Gamma_l{}^0)P(\Gamma_m{}^0) \ldots \exp[i(R + S)] \\
&\quad\times \exp\left[-i\left(e_a \int_{\Gamma_{1'1}} B^i \, da_i\right.\right. \\
&\quad\left.\left. + e_b \int_{\Gamma_{2'2}} B^i \, db_i + \ldots + e \oint_{\Gamma_l{}^0} B^i \, dl_i + e \oint_{\Gamma_m{}^0} B^i \, dm_i + \ldots\right)\right] \\
&\quad\cdot \mathscr{D}^3\Gamma_{1'1}\, \mathscr{D}^3\Gamma_{2'2} \ldots \mathscr{D}^3\Gamma_l{}^0\, \mathscr{D}^3\Gamma_m{}^0 \ldots .
\end{aligned} \tag{125}$$

The factor $\exp(iR)$ is the same as in §A3.2. The factor $\exp(iS)$ is new and gives the interactions between particles and loops and between loops and loops, through

$$S = -\sum_a \sum_l e_a e \int\int \delta(q^2{}_{AL})\, da_i\, dl^i - \frac{1}{2}e^2 \sum_l \sum_m \int\int \delta(q^2{}_{LM})\, dl^i\, dm_i. \tag{126}$$

The many-particle propagator is given by summing (125) for $n = 0, 1, 2, \ldots$. The factor $\frac{1}{2}$ appears in the second term of (126) because the pair of loops (l, m) is counted twice. The particle a and the loop l are only counted once in the first term. In the above formulas e is the electron charge, which implies that we are considering the electron vacuum, although even powers of e always appear in practical calculations, so the sign of e is irrelevant.

The discussion now proceeds as it did in §A3.2. For every pair of particles we had paths $\boldsymbol{a}$, $\boldsymbol{b}$, and conjugate paths $\boldsymbol{a}'$, $\boldsymbol{b}'$, leading to the contributions $F[\boldsymbol{a}, \boldsymbol{b}; \boldsymbol{a}', \boldsymbol{b}']$, $F[\boldsymbol{a}, \boldsymbol{a}']$, $F[\boldsymbol{b}, \boldsymbol{b}']$, to the influence functional. For every pairs of loops (l, m) we now have closed paths $\boldsymbol{l}$, $\boldsymbol{m}$, and closed conjugate paths $\boldsymbol{l}'$, $\boldsymbol{m}'$, giving contributions $F[\boldsymbol{l}, \boldsymbol{m}; \boldsymbol{l}', \boldsymbol{m}']$, $F[\boldsymbol{l}, \boldsymbol{l}']$, $F[\boldsymbol{m}, \boldsymbol{m}']$. These lead to vacuum diagrams—they have the effect of multiplying amplitudes by a phase factor $\exp iC$, C real. Such factors disappear in the present theory because these contributions to the influence functional appear only after the response of the universe has been used, and we are then concerned with the calculation of a probability, not an amplitude. Loops by themselves are of no relevance. But the mixed contributions $F[\boldsymbol{a}, \boldsymbol{l}; \boldsymbol{a}', \boldsymbol{l}']$ are of physical importance. They produce a scattering of particle $\boldsymbol{a}$ and contribute $-27 Mc/\mathrm{sec}$ to the Lamb shift. Using the method of §A3.2 we obtain

$$F[\boldsymbol{a}, \boldsymbol{l}; \boldsymbol{a}', \boldsymbol{l}']$$

$$= \exp\Big\{iee_a\Big[\iint_{t_{A'} > t_L} \delta_+(q^2{}_{A'L})\, da'^i\, dl_i - \iint_{t_L > t_{A'}} \delta_-(q^2{}_{LA'})\, da'^i\, dl_i$$

$$+ \iint_{t_{L'} > t_A} \delta_+(q^2{}_{L'A})\, da^i\, dl_i' - \iint_{t_A > t_{L'}} \delta_-(q^2{}_{AL'})\, da^i\, dl_i'$$

$$- \iint \delta_+(q^2{}_{AL})\, da^i\, dl_i + \iint \delta_-(q^2{}_{A'L'})\, da'^i\, dl_i'\Big]\Big\}. \tag{127}$$

This completes the program set out at the beginning of this section. We have now obtained the required influence functionals between particles and particles, between particles and loops, and the influence functional representing the action of a particle on itself—the radiation reaction. All the usual results of quantum electrodynamics follow by applying the perturbation expansion to these influence functionals.

We are now in an equivocal position, however, for just as we can obtain all the usual correct practical results of quantum electrodynamics, so we can obtain all the unwanted divergences in radiation correction processes. The critical question we now have to ask is whether these divergences are inevitable, or does the present direct-particle theory provide an escape from them, as it does in classical theory? We shall consider this question in the next two sections.

§A3.4. The Electron Self-Energy and the Radiative Correction to Scattering

We come now to the analogue of the self-energy divergence in field theory. For a particle we have

$$-\frac{1}{2}e_a{}^2\int\int \delta(q^2{}_{A\tilde{A}})\,da^i\,d\tilde{a}_i, \tag{128}$$

which together with the response of the universe leads to the term

$$-\frac{1}{2}e_a{}^2\int\int \delta_+(q^2{}_{A\tilde{A}})\,da^i\,d\tilde{a}_i \tag{129}$$

in the influence functional associated with particle a. We discussed the interpretation of (128) at the end of §A3.2, and saw that to avoid ambiguity in the meaning of the line integrals points A and $\tilde{A}$ should be distinct. In §A3.1 we defined the path steps by a set of points X_i, $i = 0, 1, \ldots, N$. We left open the question of whether N should tend to infinity. If we decide affirmatively and arrange X_i so that all steps tend to zero, $A \to \tilde{A}$ cannot be avoided and the outcome is completely equivalent to the usual theory. We obtain all the usual divergences and require all the usual apparatus of renormalization theory. Here, however, we propose to examine the effect of preventing $A \to \tilde{A}$. To preserve a four-dimensional notation we require

$$|X_A{}^i - X_{\tilde{A}}{}^i| \geq |\epsilon^i|, \qquad i = 1, 2, 3, 4, \tag{130}$$

where the $|\epsilon^i|$ are to be regarded as small compared to the Compton wavelength. To make (130) physically meaningful, ϵ^i must be a vector associated with the particle. For a free particle we can try

$$\epsilon^i = \epsilon \cdot \frac{p^i}{m}, \tag{131}$$

where ϵ is to be a length very small compared to m^{-1}. Thus we can choose

$$\epsilon = 2Gm, \tag{132}$$

the gravitational radius of the particle, giving $m\epsilon \approx 10^{-40}$. Although we have not permitted the path steps to tend strictly to zero, the scale of the steps is nevertheless exceedingly small compared to any physical structures that have hitherto been considered.

We emphasize that we are not seeking here to justify (130), (131), (132), but to investigate their effect and to compare the resulting situation with the cutoff procedures of the usual theory. If indeed gravitation plays a critical role, then modification of the spacetime structure on the scale ϵ is required. Only through such an investigation could one expect to understand ϵ^i more deeply.

We show in this section that (130), (131), (132), lead to a finite self-energy even though the frequency in the three-dimensional Fourier integral of the $\delta_+(q^2)$ function tends to infinity. And in §A3.5 we shall show that the divergences in vacuum polarization, which are usually thought to be different in nature from the divergence of the self-energy, are removed by the same procedure.

For simplicity of presentation we calculate first the self-energy for an electron with $\boldsymbol{p} = \boldsymbol{0}$. Suppose at $t = 0$ we are given the wave function $\mathscr{U}_0$ corresponding to positive energy m. We wish to examine the effect of the term (129) in the influence functional for the slab $0 \leq t \leq T$. The wave function at point 2 on $t = T$ is usually calculated from the perturbation expansion

$$\mathscr{U}_0 e^{-imT}$$

$$- ie^2 \int\int K_+(2,3)\gamma_i K_+(3,4)\gamma^i \, \delta_+(q^2{}_{34})\mathscr{U}_0 e^{-imt_4} \, d\tau_3 \, d\tau_4 + \ldots, \quad (133)$$

where the integrations with respect to $d\tau_3$, $d\tau_4$, are over the slab. This permits points 3 and 4 to coincide, which is the situation we now wish to avoid. From (130) and (131) with $p^i = (-\boldsymbol{p}, E) = (0, m)$, the evaluation of (133) is now subject to

$$|t_3 - t_4| \geq \epsilon. \quad (134)$$

To deal with the restriction (134) we separate the second term of (133) into two terms Q_1, Q_2, such that $t_3 > t_4$ in Q_1 and $t_3 < t_4$ in Q_2.

For $T \geq t_3 > t_4 \geq 0$ write

$$K_+(2,3)$$

$$= \int_{E_2 > 0} \frac{1}{2E_2} (\gamma_4 E_2 - \boldsymbol{\gamma} \cdot \boldsymbol{p}_2 + m) e^{-iE_2(T - t_3) + i\boldsymbol{p}_2 \cdot (\boldsymbol{x}_2 - \boldsymbol{x}_3)} \frac{d^3 p_3}{(2\pi)^3}, \quad (135)$$

$$K_+(3;4)$$

$$= \int_{E > 0} \frac{1}{2E} (\gamma_4 E - \boldsymbol{\gamma} \cdot \boldsymbol{p} + m) e^{-iE(t_3 - t_4) + i\boldsymbol{p} \cdot (\boldsymbol{x}_3 - \boldsymbol{x}_4)} \frac{d^3 p}{(2\pi)^3}, \quad (136)$$

where

$$E_2^2 = \boldsymbol{p}_2^2 + m^2, \qquad E^2 = \boldsymbol{p}^2 + m^2. \tag{137}$$

The momentum representation for $\delta_+(q^2_{34})$ is given in (102). Performing integrations with respect to $d^3\boldsymbol{x}_3$, $d^3\boldsymbol{x}_4$, $d^3\boldsymbol{p}$, $d^3\boldsymbol{p}_2$, we get

$$\boldsymbol{p}_2 = 0, \qquad \boldsymbol{p} = -\boldsymbol{k}, \tag{138}$$

so that $E_2 = m$, $E = \sqrt{K^2 + m^2}$, and Q_1 becomes

$$Q_1 = (-ie^2)\cdot(4\pi i)\cdot\frac{1}{2^3}\int\int\int_{t_3>t_4}(\gamma_4 + 1)\gamma_i(\gamma_4 E + \gamma\cdot\boldsymbol{k} + m)\gamma^i\mathscr{U}_0\cdot$$
$$\frac{1}{EK}\cdot e^{-imT+i(m-E-K)(t_3-t_4)}\frac{d^3\boldsymbol{k}}{(2\pi)^3}\,dt_3\,dt_4. \tag{139}$$

Using symmetrical integration to remove terms linear in $\boldsymbol{k}$, Q_1 is simplified to

$$Q_1 = \xi_1\mathscr{U}_0 e^{-imT}, \tag{140}$$

where

$$\xi_1 = \frac{e^2}{\pi}\int\int\int_{t_3>t_4}\frac{2m - \sqrt{K^2+m^2}}{\sqrt{K^2+m^2}}\,e^{i(m-E-K)(t_3-t_4)}K\,dK\,dt_3\,dt_4. \tag{141}$$

Similarly, we have

$$Q_2 = \xi_2\mathscr{U}_0 e^{-imT}, \tag{142}$$

where

$$\xi_2 = \frac{e^2}{\pi}\int\int\int_{t_4>t_3}\frac{2m + \sqrt{K^2+m^2}}{\sqrt{K^2+m^2}}\,e^{i(m+E+K)(t_3-t_4)}K\,dK\,dt_3\,dt_4. \tag{143}$$

It is usually argued that divergences in Q_1, Q_2 arise from high-frequency quanta, i.e., when $K\to\infty$. This is based on a calculation in which the integrations with respect to time are performed first. Now because of the time restrictions we integrate first with respect to frequency. For $t_3 \neq t_4$ the integrals with respect to K converge. To see this we examine the behavior of (141) and (143) at large K. Thus for $K \gg m$,

$$\xi_1 \sim \frac{e^2}{\pi} \iiint_{t_3 > t_4} (2m - K) e^{i(m-2K)(t_3 - t_4)} \, dK \, dt_3 \, dt_4$$

$$= \frac{e^2}{\pi} \iint_{t_3 > t_4} e^{im(t_3 - t_4)} \int_0^\infty (2m - K) e^{-2iK(t_3 - t_4)} \, dK \, dt_3 \, dt_4, \quad (144)$$

and

$$\xi_2 \sim \frac{e^2}{\pi} \iint_{t_3 < t_4} e^{im(t_3 - t_4)} \int_0^\infty (2m + K) e^{2iK(t_3 - t_4)} \, dK \, dt_3 \, dt_4. \quad (145)$$

The actual values of ξ_1, ξ_2, differ from these asymptotic values by finite quantities.

Equations (144) and (145) can be simplified further by using the results

$$\lambda \neq 0, \qquad \int_0^\infty e^{-i\lambda x} \, dx = \frac{1}{i\lambda}, \qquad \int_0^\infty x e^{-i\lambda x} \, dx = -\frac{1}{\lambda^2}. \quad (146)$$

Thus

$$\xi_1 \sim \frac{e^2}{\pi} \iint_{t_3 > t_4} e^{im(t_3 - t_4)} \left[\frac{-im}{t_3 - t_4} + \frac{1}{4(t_3 - t_4)^2} \right] dt_3 \, dt_4, \quad (147)$$

$$\xi_2 \sim \frac{e^2}{\pi} \iint_{t_3 < t_4} e^{im(t_3 - t_4)} \left[\frac{im}{t_3 - t_4} - \frac{1}{4(t_3 - t_4)^2} \right] dt_3 \, dt_4. \quad (148)$$

Interchanging the naming of t_3 and t_4 in ξ_2 we have

$$\xi_1 + \xi_2 \sim \frac{2e^2}{\pi} \iint_{t_3 > t_4} \left[-im \frac{\cos m(t_3 - t_4)}{t_3 - t_4} + i \frac{\sin m(t_3 - t_4)}{4(t_3 - t_4)^2} \right] dt_3 \, dt_4$$

$$= \frac{2ie^2}{\pi} \int_0^{mT} dx \int_{m\epsilon}^{x} \left[-\frac{\cos y}{y} + \frac{\sin y}{4y^2} \right] dy, \quad (149)$$

where

$$x = mt_3, \qquad y = m(t_3 - t_4), \quad (150)$$

and we have now taken account of the restriction $|t_3 - t_4| \geq \epsilon$.

For a slab with $mT \gg 1$, which is the situation usually considered, we have

$$\xi_1 + \xi_2 \sim \frac{2ie^2}{\pi} \int_0^{mT} dx \int_{m\epsilon}^\infty \frac{1}{y} \left(\frac{\sin y}{4y} - \cos y \right) dy. \quad (151)$$

For $m\epsilon \ll 1$, the main contribution to (151) comes from $y \ll 1$ when the trigonometrical factor $\sim -\frac{3}{4}$, and

$$\xi_1 + \xi_2 \sim -iAT, \tag{152}$$

where

$$A = -\frac{3e^2}{2\pi} m \ln(m\epsilon). \tag{153}$$

Since we are considering paths to have finite time steps ϵ, it would be a stricter procedure to divide the whole time-range $[0, T]$ into many steps of length ϵ and to work out $\xi_1 + \xi_2$ as a sum over steps rather than as an integral. It turns out when this is done that the result is the same as (153).

The wave function at $t = T$ to order e^2 is therefore given by

$$\mathcal{U}_0 e^{-imT}\left[1 + i \cdot \frac{3e^2}{2\pi} m \ln(m\epsilon)\, T\right] \approx \mathcal{U}_0 e^{-im'T} \tag{154}$$

for T not too large, where

$$m' = m\left[1 - \frac{3e^2}{2\pi} \ln(m\epsilon)\right] = m + A. \tag{155}$$

The effect of the response of the universe is to "change" the mass m to m' given by (155). The usual theory gives

$$A = \frac{3e^2}{2\pi} m \ln \frac{K_0}{m}, \tag{156}$$

where K_0 is a cutoff applied to the K integration, which is carried out after integrations with respect to t_3, t_4. Although (153) and (156) are similar in form, the cutoff in (153) can be given a physical basis, for example, in terms of the gravitational radius of the particle, whereas that in (156) is a mathematical variable from a Fourier integral and should not play a role in the theory.

We wish next to consider the corresponding problem for a free particle with $p^i = (-\boldsymbol{p}, +E)$, $\boldsymbol{p} \neq 0$, $E > 0$, and 4-spinor $\mathcal{U}$. According to (130), (131), there are restrictions now on all the coordinates. This leads to a more complicated situation. It is possible, however, to simplify the work by noting first that if, as before, we restrict only the time-coordinate, by $|t_3 - t_4| \geq \epsilon^4$, the outcome for the wave function at point 2 on $t = T$ is

$$\mathcal{U} \exp[i\boldsymbol{p} \cdot \boldsymbol{x}_2 - i(E + \Delta E)T], \tag{157}$$

where[2]

$$E\,\Delta E = -\frac{3e^2m^2}{2\pi}\ln(E\epsilon^4). \tag{158}$$

Since we are integrating invariant functions over a four-dimensional region, it is clear that if we had used

$$|X_3^{\mu} - X_4^{\mu}| \geq \epsilon^{\mu} \tag{159}$$

to restrict one of the space coordinates, $\mu = 1, 2, 3$, instead of the time coordinate, we should have obtained $\ln(p_\mu\epsilon^\mu)$ instead of $\ln(E\epsilon^4)$, and that if we had used

$$|X_3^{i} - X_4^{i}| \geq \epsilon^{i}, \qquad i = 1, 2, 3, 4, \tag{160}$$

for all coordinates we should have obtained $\ln(p_i\epsilon^i)$. This is simply $\ln(m\epsilon)$, so that in place of (158) we have the invariant result

$$E\,\Delta E = -\frac{3e^2}{2\pi}m^2\ln(m\epsilon) = m\,\Delta m. \tag{161}$$

Unlike the usual theory the so-called "wave function renormalization," represented by the constant

$$B = -\frac{e^2}{2\pi}\ln(m\epsilon), \tag{162}$$

did not appear in the above work. The constant B does appear, however, if we seek the wave function at an intermediate time t, $0 < t < T$. We again consider the case of a particle with wave function $\mathcal{U}_0$ corresponding to energy m at $t = 0$. If we were to restrict our time-integrals to the interval $(0, t)$ the analysis would be the same as that given above, but with t replacing T, and we should obtain $\mathcal{U}_0 e^{-im't}$ for the wave function at time t. But we must permit integrations up to T. This does not change Q_1, which has $t \geq t_3 > t_4 \geq 0$. The restriction $t_3 \leq t$ here arises because $\delta_+(q^2{}_{34})$ leaves the particle in a positive-energy state after the response is completed at point 3, and hence we must go forward in time from t_3 to t. But for Q_2, in addition to $t \geq t_4 > t_3 \geq 0$, we have also to consider

$$T \geq t_4 > t > t_3 \geq 0. \tag{163}$$

[2]The proof of (158) is contained in Hoyle and Narlikar (1971).

It is this additional time-range that makes a contribution to the wave function at time t, over and above $\mathcal{U}_0 e^{-im't}$.

The extra contribution is

$$\xi_2' \mathcal{U}_0 e^{-imt}, \tag{164}$$

where

$$\begin{aligned}\xi_2' &\sim \frac{e^2}{\pi} \iint_{T \geq t_4 > t > t_3 \geq 0} e^{im(t_3 - t_4)} \int (2m + K) e^{2iK(t_3 - t_4)}\, dK\, dt_3\, dt_4 \\ &\sim \frac{e^2}{\pi} \iint_{T \geq t_4 > t > t_3 \geq 0} e^{im(t_3 - t_4)} \left[\frac{-im}{t_4 - t_3} - \frac{1}{4(t_4 - t_3)^2} \right] dt_3\, dt_4. \end{aligned} \tag{165}$$

The intervals for t_3, t_4, have been kept open at t in the above formulas to indicate that $t_4 - t_3$ has a minimum value ϵ. We require

$$t_3 \leq t - \alpha\epsilon, \qquad t_4 \geq t + \beta\epsilon, \tag{166}$$

where α, β, are positive numbers satisfying $\alpha + \beta = 1$. All possible choices for α, β, lead to the same final result. Taking $\alpha = 0$, and making the substitutions

$$x = m(t_4 - t), \qquad y = m(t_4 - t_3), \tag{167}$$

we obtain

$$\xi_2' \sim -\frac{e^2}{\pi} \int_{m\epsilon}^{m(T-t)} dx \int_x^{x+mt} e^{-iy} \left(\frac{i}{y} + \frac{1}{4y^2} \right) dy. \tag{168}$$

Provided t is not close to zero or to T, the $\frac{1}{4}y^2$ in (168) gives a large contribution when both x and y are close to $m\epsilon$. After integrating with respect to y for this term we get

$$\xi_2' \sim -\frac{e^2}{4\pi} \int_{m\epsilon} \frac{dx}{x} \sim \frac{e^2}{4\pi} \ln(m\epsilon) \sim -\frac{1}{2} B. \tag{169}$$

Provided t is not close to zero or to T, the required wave function to order e^2 is therefore

$$\left(1 - iAT - \frac{1}{2} B \right) \mathcal{U}_0 e^{-imt}. \tag{170}$$

The appearance of B within the slab is of importance when an external potential acts on the particle. In general an external potential[3] can be represented by a four-dimensional Fourier integral

$$B_i(X) = \int b_i(q) e^{-iq \cdot x} \frac{d^4 q}{(2\pi)^4}. \tag{171}$$

The scattering problem is to determine the amplitude for a state of momentum $\boldsymbol{p}_2$ at $t_2 = T$ given that the state has momentum $\boldsymbol{p}_1$ at $t_1 = 0$. From the path-integral point of view we have to consider the effect of the phase factor

$$\exp\left(-ie \int_{\Gamma_{21}} B_i \, dx^i\right) \tag{172}$$

on this problem.

Expanding (172) as a power series, the unity term leads to (154), which involves no scattering. The linear term in B_i produces scattering, however. In most cases of physical interest, higher-order terms in B_i give smaller contributions to the scattering than the linear term; so it is usual to restrict the problem to determining the effect of the linear term. It is found that only those components in the Fourier integral for which

$$\boldsymbol{q} \cong \boldsymbol{p}_2 - \boldsymbol{p}_1, \qquad q_4 \cong E_2 - E_1, \tag{173}$$

where $E_2 = (\boldsymbol{p}_2{}^2 + m^2)^{1/2}$, $E_1 = (\boldsymbol{p}_1{}^2 + m^2)^{1/2}$, make a contribution to the scattering into states of momentum $\boldsymbol{p}_2$. For finite T the Fourier components making effective contribution can depart from the strict equalities of (173) by small terms of order T^{-1}.

The scattering amplitude can be developed in a series of ascending powers of q^i. Since b^i, q^i, appear in relativistically invariant forms, and since $b(q) \cdot q = 0$ because of the Lorentz condition, the b^i components can only appear in the combination $\not{b}$, while the q^i components can appear as q^{2n+2} or as $q^{2n}\not{q}$, $n = 0, 1, 2, \ldots$. The most general form for the scattering amplitude is therefore seen to be

$$\not{b}(q) \sum_{n=0}^{\infty} C_n q^{2n} + [\not{b}(q)\not{q} - \not{q}\not{b}(q)] \sum_{n=0}^{\infty} D_n q^{2n}, \tag{174}$$

[3]The reader will note the clash between our notation B_i for the external potential and B for the constant $-e^2 \ln (m\epsilon)/2\pi$.

where the C_n, D_n, are constants. Detailed calculations in quantum electrodynamics show that none of C_n, $n = 1, 2, \ldots$, or D_n, $n = 0, 1, 2, \ldots$, involve the cutoff $\ln(K_0/m)$. Because the present theory differs from the usual theory only in the form of the cutoff, not in the places where it occurs, the present theory leads to (174) with the same coefficients—except possibly for C_0. It is well-known that the second series in (174) gives the anomalous magnetic moment of the electron, and the entire series without the C_0 term gives the Lamb shift. The observable effects of including

$$\exp\left[-\frac{1}{2}e^2 \int\int \delta_+(q^2{}_{A\tilde{A}})\, da^i\, d\tilde{a}_i\right] \tag{175}$$

in scattering are therefore the same in this theory as they are in the usual theory.

Although we are working only to the first order in b_i, it is important to notice that (174) includes all orders in the expansion of (175)—in the usual language we can have any number of virtual photons. Each order in e^2 is handled by Feynman–Dyson graphs. It is found that while individual graphs give cutoff-dependent contributions to C_0, the sum of the contributions of all graphs does not involve the cutoff. The well-known situation to order e^2 is shown in Figure A3.1. The cutoff-dependent term from I cancels those from II and III.

Once again we expect the same result for C_0 in this theory. Here we use a method depending on the closed form of the electromagnetic phase factor in the path integrals, to show that C_0 must be cutoff-independent to all orders in e^2, and hence it follows that the detailed calculations based on Feynman–Dyson graphs must lead to the cancelations described above.

The C_0 term in (174) survives as the components q^i tend to zero. Let the field B_i be such that $b_i(q)$ is zero except near $q^i = 0$ for all i: that is to

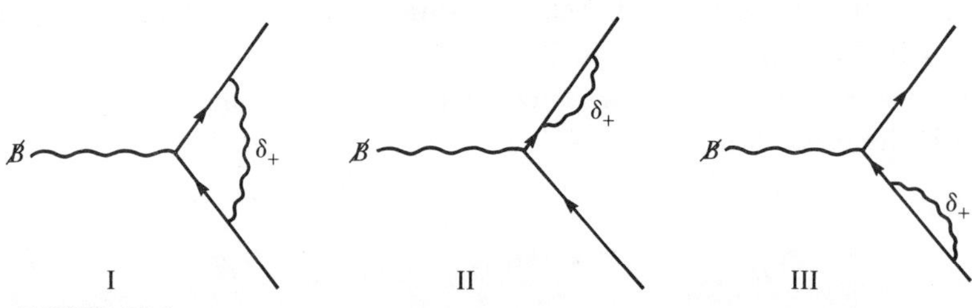

FIGURE A3.1
The three situations that arise when the external potential and the response of the universe act once in the perturbation expansion. The infinite B-term in I is exactly canceled by similar terms arising from II and III.

say, we collapse the Fourier coefficients b^i on to the origin in q-space. As the region in which $b_i(q)$ is nonzero shrinks, B_i becomes effectively constant, and

$$\exp\left(-ie\int_{\Gamma_{21}} B_i\,dx^i\right) \to \exp[-ie\boldsymbol{B}\cdot(\boldsymbol{x}_2 - \boldsymbol{x}_1)]. \tag{176}$$

The electromagnetic phase factor arising from B_i becomes path-independent. The gauge transformation

$$B_i \to B_i - \frac{\partial\chi}{\partial x^i}, \qquad \chi = B_i x^i, \tag{177}$$

reduces the field to zero. Since a gauge transformation produces no physical effect, the situation is the same as for a zero field. There can be no scattering and therefore no cutoff-dependent terms in C_0.

Consider the case $B^i = (-\boldsymbol{B}, 0)$ in further detail, writing $\boldsymbol{p} = -e\boldsymbol{B}$. Let the four-component spinor at $t = 0$ again be $\mathscr{U}_0$ corresponding to a free-particle state of energy m. Suppose that the wave function on $t = 0$ has no spatial dependence. Then we have to consider $\mathscr{U}_0 e^{i\boldsymbol{p}\cdot\boldsymbol{x}_1}$ on $t = 0$ because of the $e^{i\boldsymbol{p}\cdot\boldsymbol{x}_1}$ factor in (176). Before propagation to $t = T$ the product $\mathscr{U}_0 e^{i\boldsymbol{p}\cdot\boldsymbol{x}_1}$ must be separated into positive-energy and negative-energy components for $\pm E = \pm\sqrt{p^2 + m^2}$. Write

$$\mathscr{U}_0 = \mathscr{U}_+ + \mathscr{U}_-, \tag{178}$$

where $\exp[i(\boldsymbol{p}\cdot\boldsymbol{x} \mp Et)]\mathscr{U}_\pm$ are solutions of the free-particle Dirac equation. Only $\mathscr{U}_+$ is propagated forward to $t = T$. Propagation subject to (175) was studied earlier in this section. After multiplying by $e^{-i\boldsymbol{p}\cdot\boldsymbol{x}_2}$ from (176) the wave function at point 2 on $t = T$ is

$$\mathscr{U}_+ e^{-i(E + \Delta E)T}, \tag{179}$$

where ΔE is given by $E\,\Delta E = m\,\Delta m$, ΔE being defined in (161). If the field had been zero, the wave function on $t = T$ would have been

$$\mathscr{U}_0 e^{-i(m + \Delta m)T}. \tag{180}$$

The difference between (179) and (180) appears to contradict what was said in the previous paragraph. The difference has arisen from the specification of the wave function at $t = 0$. Propagation from $t < 0$ cannot give a wave function on $t = 0$ with spinor $\mathscr{U}_0$ and without an $\boldsymbol{x}_1$ dependence

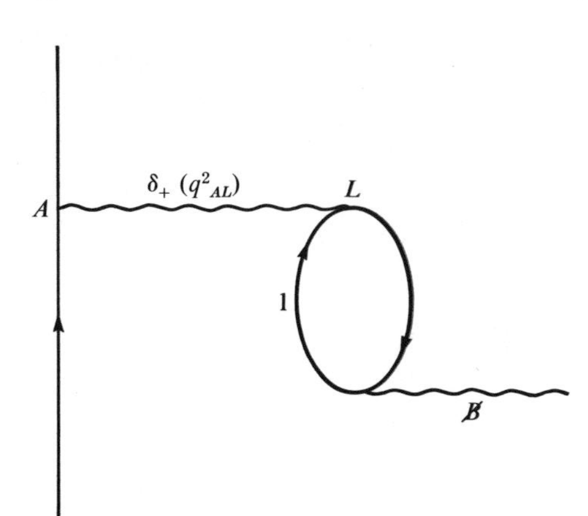

FIGURE A3.2
The simplest vacuum diagram leading to an observable result (-27 Mc/sec in the Lamb shift).

unless $B_i = 0$ for $t < 0$. We have therefore introduced a discontinuity in the field at $t = 0$, a discontinuity that cannot be removed by a gauge transformation. Indeed, the discontinuity generates components in the Fourier integral (171) for b_i away from the origin in q-space, which is just the situation we set out to avoid. It is therefore necessary, if we wish to have no x_1^- dependence on $t = 0$ to specify $\mathcal{U}_+$ as the 4-spinor. Alternatively, if we wish to specify $\mathcal{U}_0$ we must have an x_1 dependence on $t = 0$ given by the factor $e^{-ip\cdot x_1}$. In both these cases the constant field has no effect.[4]

§A3.5. Vacuum Polarization

The amplitude for the process shown in Figure A3.2 is

$$-(-ie^2)(-ie)\int\int\int \bar{u}_2(5)\gamma_i u_1(5)$$
$$\delta_+(q^2{}_{54})\,\mathrm{Tr}[\gamma^i K_+(4;3)B(3)K_+(3;4)]\,d\tau_3\,d\tau_4\,d\tau_5, \quad (181)$$

the integrations with respect to $d\tau_3$, $d\tau_4$, $d\tau_5$, being over the slab $0 \le t \le T$.

[4] It is well-known that an invariant finite cut-off procedure, such as we suggest here, is not consistent with a strict conservation of probabilities. Finite radiation processes involving convergent integrals are cutoff at ϵ^{-1} instead of going to infinity with respect to frequency. This leads to a very small change of order $m\epsilon$ in the calculated probabilities. The question of conservation of probabilities arises in this order. Our point of view is that this question cannot be discussed within the framework of Minkowski spacetime. The gravitational effect of an electron distorts spacetime to the same order as the question at issue.

While this paper was in press, one expressing a similar point of view by Salam and Strathdee (1970) appeared. They discuss this same issue at greater length and in more detail than we have done.

The expression (181) is the lowest-order contribution of the vacuum in the presence of an external field B_i to the scattering of a particle from state u_1 into state u_2. This amplitude follows without difficulty from the general path integral of §A3.3. Since the integral of a constant potential around a closed path is zero, it is always possible to subtract an arbitrary constant potential from B_i. We shall make use of this possibility at a later stage.

The scattering is the same as that produced by a potential

$$A^i(5) = ie^2 \int\int \delta_+(q^2{}_{54}) \operatorname{Tr}[\gamma^i K_+(4;3) \not{B}(3) K_+(3;4)]\, d\tau_3\, d\tau_4. \qquad (182)$$

This is the potential produced by a current $j^i(5)$ given by

$$j^i(5) = \frac{1}{4\pi} \Box_5 A^i(5) = ie^2 \int \operatorname{Tr}[\gamma^i K_+(5;3) \not{B}(3) K_+(3;5)]\, d\tau_3. \qquad (183)$$

In this section we show that (183) can be evaluated in such a way that no quadratic divergence is present, and we show that the cutoff-dependent logarithmic term is of the same form as that encountered in the preceding section. Given j^i we can construct A^i and hence obtain the scattering from

$$-ie \int \bar{u}_2(5) \not{A}(5) u_1(5)\, d\tau_5. \qquad (184)$$

With a slight change of notation we write

$$j_k(1) = ie^2 \int \operatorname{Tr}[K_+(1;2)(\not{B}_2 - \not{B}_1) K_+(2;1) \gamma_k]\, d\tau_2 \qquad (185)$$

where $\not{B}_2$, $\not{B}_1$, are the values of $\not{B}$ at points 2 and 1. The potential B_1 is constant and we are now making use of the possibility of subtracting a constant potential from the variable potential $\not{B}_2$. This step is useful in respect of the Taylor expansion

$$\not{B}_2 - \not{B}_1 = \gamma^i \left(B_{i,l} q^l + \frac{1}{2} B_{i,lm} q^l q^m + \dots \right), \qquad (186)$$

where the derivatives $B_{i,l}$, $B_{i,lm}$, are evaluated at point 1, and $q^i = x_2{}^i - x_1{}^i$. We write the propagators in the form

$$K_+(2;1) = (-\not{\nabla}_2 + im) I_+(q^2) = -2\not{q} I_+'(q^2) + im I_+(q^2), \qquad (187)$$

$$K_+(1;2) = 2\not{q} I_+'(q^2) + im I_+(q^2), \qquad (188)$$

where

$$I_+'(q^2) = \frac{d}{dq^2} I_+(q^2), \tag{189}$$

and

$$I_+(q^2) = -\frac{1}{4\pi}\delta(q^2) + \frac{m}{8\pi q} H_1^{(2)}(mq). \tag{190}$$

In the Hankel function q is the positive square root of q^2 when $q^2 > 0$ and is the negative square root when $q^2 < 0$. The limiting form of $I_+(q^2)$ as $q^2 \rightarrow 0$ is

$$I_+(q^2) \rightarrow -\frac{1}{4\pi}\delta_+(q^2).$$

Inserting (186), (187), (188) in (185) we get

$$j_k(1) = ie^2 \int \mathrm{Tr}[(2\gamma^p q_p I_+' + imI_+)\left(B_{i,l}q^l + \frac{1}{2}B_{i,lm}q^l q^m + \ldots\right)\gamma^i$$
$$\cdot(-2\gamma^r q_r I_+' + imI_+)\gamma_k]\, d^4q. \tag{191}$$

To simplify (191) we note that the integrations remove terms of odd order in q_l, and that the trace removes terms that contain an odd number of γ matrices. Thus

$$j_k(1) = (ie^2)\cdot B_{i,lm}\int \mathrm{Tr}\Big(-2\gamma^p\gamma^i\gamma^r\gamma_k q_p q^l q^m q_r I_+'^2$$
$$-\frac{1}{2}m^2 I_+^2\gamma^i\gamma_k q^l q^m + \ldots\Big)\, d^4q. \tag{192}$$

Next we use

$$\mathrm{Tr}(\gamma^i\gamma_k) = 4\eta^i{}_k,$$
$$\mathrm{Tr}(\gamma^p\gamma^i\gamma^r\gamma_k) = 4(\eta^{pi}\eta^r{}_k + \eta^p{}_k\eta^{ir} - \eta^{pr}\eta^i{}_k), \tag{193}$$

to give

$$j_k(1) = -8ie^2 B_{i,lm}\int (2q^i q_k - q^2\eta^i{}_k) I_+'^2 q^l q^m\, d^4q$$
$$-2ie^2 m^2 B_{i,lm}\int \eta^i{}_k I_+^2 q^l q^m\, d^4q + \ldots. \tag{194}$$

Only the first term in (194) is cutoff-dependent. The rest remain finite as $\epsilon \to 0$. Since it is the cutoff-dependent term we are seeking, we shall confine further attention to this first term.

By symmetry we can write

$$\int I_+'^2 q^l q^m q^i q^k \, d^4q = \lambda(\eta^{lm}\eta^{ik} + \eta^{li}\eta^{mk} + \eta^{lk}\eta^{mi}). \tag{195}$$

Contracting l against m and i against k gives

$$\int \{I_+'(q^2)q^2\}^2 \, d^4q = 24\lambda. \tag{196}$$

The integral on the lefthand side of this equation is worked out in Hoyle and Narlikar (1971). Using the results there, we have

$$\lambda = -\frac{i}{192\pi^2}\ln(m\epsilon). \tag{197}$$

Returning to the first term of (194) we note that

$$\begin{aligned} B_{i,\,lm}\int (2q^i q_k - q^2\eta^i{}_k) I_+'^2 q^l q^m \, d^4q \\ &= 2\lambda B_{i,\,lm}(\eta^{lm}\eta^i{}_k + \eta^{li}\eta^m{}_k + \eta^l{}_k\eta^{mi} - 3\eta^{lm}\eta^i{}_k) \\ &= -4\lambda \Box B_k + 2\lambda B^l{}_{,lk} + 2\lambda B^m{}_{,km} \\ &= -4\lambda \Box B_k, \end{aligned} \tag{198}$$

since $B^m{}_{,km} = B^m{}_{,mk} = 0$ and $B^l{}_{,lk} = 0$. Using (198) we get

$$j_k(1) \sim 32ie^2\lambda \Box B_k. \tag{199}$$

With (197) for λ,

$$j_k(1) \sim \frac{e^2}{6\pi^2}\ln(m\epsilon)\Box B_k. \tag{200}$$

The d'Alembertian of B_k is taken at point 1. Hence there is no vacuum current except at the sources of the external field B_k. In the direct-particle theory this means there is no polarization current except at the particles. If J_k is the particle current, we have $\Box B_k = 4\pi J_k$ and

$$j_k(1) = \frac{2e^2}{3\pi}\ln(m\epsilon)J_k(1). \tag{201}$$

The result that the polarization current occurs only at the particles holds also for the finite terms not considered explicitly above.

The usual theory gives

$$j_k(1) \sim -\frac{2e^2}{3\pi} \ln\left(\frac{M}{m}\right) J_k(1), \tag{202}$$

in terms of a momentum cutoff M. In §A3.4 we saw that ϵ in this theory replaces the photon cutoff K_0 of the usual theory. Now we see that ϵ also replaced the momentum cutoff of the usual theory. There is no requirement for two distinct modes of cutoff.

We also note that the quadratic divergence, encountered in most treatments of the usual theory, did not appear in the above work (see Hoyle and Narlikar, 1971).

APPENDIX 4 (a note to §6.3)

We are required to show that $m^* = \Omega^{-1}m$, $\psi^* = \Omega^{-3/2}\psi$, makes the Dirac equation conformally invariant. The proof of this important result was deferred in the main text because the work of §8.3 is needed for it. Using the representation in which

$$\psi = \begin{bmatrix} u_\alpha \\ v^{\dot\beta} \end{bmatrix}, \tag{203}$$

the Dirac equation for unstarred quantities takes the form

$$\begin{bmatrix} 0 & g^i{}_{\alpha\dot\beta} \\ g^{i\alpha\dot\beta} & 0 \end{bmatrix} \begin{bmatrix} u_\alpha \\ v^{\dot\beta} \end{bmatrix}_{;i} + im \begin{bmatrix} u_\alpha \\ v^{\dot\beta} \end{bmatrix} = 0 \tag{204}$$

with respect to the metric

$$ds^2 = g_{ik}\, dx^i\, dx^k. \tag{205}$$

A conformal transformation to

$$ds^{*2} = \Omega^2 g_{ik}\, dx^i\, dx^k \tag{206}$$

is accompanied by

$$g^*{}_{i\alpha\dot\beta} = \Omega g_{i\alpha\dot\beta}, \qquad g^{*i}{}_{\alpha\dot\beta} = \Omega^{-1} g^i{}_{\alpha\dot\beta}. \tag{207}$$

We are required to prove that

$$\begin{bmatrix} u^*{}_\alpha \\ v^{*\dot\beta} \end{bmatrix} = \Omega^{-3/2}\begin{bmatrix} u_\alpha \\ v^{\dot\beta} \end{bmatrix}, \qquad m^* = \Omega^{-1}m, \tag{208}$$

satisfy

$$\begin{bmatrix} 0 & g^{*i}{}_{\alpha\dot\beta} \\ g^{*i\alpha\dot\beta} & 0 \end{bmatrix}\begin{bmatrix} u^*{}_\alpha \\ v^{*\dot\beta} \end{bmatrix}_{;i} + im^*\begin{bmatrix} u^*{}_\alpha \\ v^{*\dot\beta} \end{bmatrix} = 0. \tag{209}$$

It will be sufficient for this purpose to show that

$$g^{*i\alpha\dot\beta}u^*{}_{\alpha;\,i} + im^*v^{*\dot\beta} = 0, \tag{210}$$

since a similar proof applies to $g^{*i}{}_{\alpha\dot\beta}v^{*\dot\beta}{}_{;i} + im^*u^*{}_\alpha$.

The spinor derivatives in curved space must be interpreted in accordance with the discussion of §8.3. Thus

$$u_{\alpha;\,i} = \frac{\partial u_\alpha}{\partial x^i} - L^\beta{}_{\alpha i}u_\beta, \tag{211}$$

where

$$L^\beta{}_{\alpha i} = \frac{1}{4}\left(g^l{}_{\alpha\dot\lambda}g_m{}^{\beta\dot\lambda}\Gamma^m{}_{il} + g_l{}^{\beta\dot\lambda}\frac{\partial}{\partial x^i}g^l{}_{\alpha\dot\lambda}\right). \tag{212}$$

For the lefthand side of (210) we have

$$\begin{aligned} &g^{*i\alpha\dot\beta}u^*{}_{\alpha;\,i} + im^*v^{*\dot\beta} \\ &\quad= \Omega^{-1}g^{i\alpha\dot\beta}\left[\frac{\partial}{\partial x^i}(\Omega^{-3/2}u_\alpha) - L^{*\beta}{}_{\alpha i}\Omega^{-3/2}u_\beta\right] + i\Omega^{-5/2}mv^{\dot\beta}, \end{aligned} \tag{213}$$

in which

$$\begin{aligned} L^{*\beta}{}_{\alpha i} &= \frac{1}{4}\left(g^{*l}{}_{\alpha\dot\lambda}g_m{}^{*\beta\dot\lambda}\Gamma^{*m}{}_{li} + g_l{}^{*\beta\dot\lambda}\frac{\partial}{\partial x^i}g^{*l}{}_{\alpha\dot\lambda}\right) \\ &= L^\beta{}_{\alpha i} - \delta^\beta{}_\alpha\frac{\partial\ln\Omega}{\partial x^i} + \frac{1}{4}g^l{}_{\alpha\dot\lambda}g^{m\beta\dot\lambda}\left(g_{lm}\frac{\partial\ln\Omega}{\partial x^i} + g_{im}\frac{\partial\ln\Omega}{\partial x^l}\right. \\ &\qquad\left. - g_{il}\frac{\partial\ln\Omega}{\partial x^m}\right). \end{aligned} \tag{214}$$

Substituting (214) in (213) gives

$$\begin{aligned}\Omega^{5/2}(g^{*i\alpha\dot{\beta}}u^*{}_{\alpha;\,i} + im^*v^{*\dot{\beta}}) \\ = g^{i\alpha\dot{\beta}}u_{\alpha;\,i} + imv^{\dot{\beta}} \\ -\frac{1}{4}g^l{}_{\alpha\dot{\lambda}}g^{m\gamma\dot{\lambda}}g^{i\alpha\dot{\beta}}u_\gamma\left(g_{lm}\frac{\partial \ln \Omega}{\partial x^i} + g_{im}\frac{\partial \ln \Omega}{\partial x^l} - g_{il}\frac{\partial \ln \Omega}{\partial x^m}\right) \\ -\frac{3}{2}g^{i\alpha\dot{\beta}}u_\alpha\frac{\partial \ln \Omega}{\partial x^i} + \delta^\beta{}_\alpha g^{i\alpha\dot{\beta}}u_\beta\frac{\partial \ln \Omega}{\partial x^i}.\end{aligned} \tag{215}$$

The terms in $\partial \ln \Omega/\partial x^i$ are easily seen to give

$$-\frac{3}{2}g^{i\alpha\dot{\beta}}u_\alpha\frac{\partial \ln \Omega}{\partial x^i}, \tag{216}$$

while the remaining terms in $\ln \Omega$ are

$$\begin{aligned}-\frac{1}{4}g^l{}_{\alpha\dot{\lambda}}g_i{}^{\gamma\dot{\lambda}}g^{i\alpha\dot{\beta}}u_\gamma\frac{\partial \ln \Omega}{\partial x^l} \\ +\frac{1}{4}g_{i\alpha\dot{\lambda}}g^{m\gamma\dot{\lambda}}g^{i\alpha\dot{\beta}}u_\gamma\frac{\partial \ln \Omega}{\partial x^m}.\end{aligned} \tag{217}$$

Using

$$g_i{}^{\gamma\dot{\lambda}}g^{i\alpha\dot{\beta}} = 2\varepsilon^{\gamma\alpha}\varepsilon^{\dot{\lambda}\dot{\beta}}, \tag{218}$$

it is easy to see that (217) becomes

$$-\frac{1}{2}g^l{}_{\alpha\dot{\lambda}}\varepsilon^{\gamma\alpha}\varepsilon^{\dot{\lambda}\dot{\beta}}u_\gamma\frac{\partial \ln \Omega}{\partial x^l} + g^{m\gamma\dot{\beta}}u_\gamma\frac{\partial \ln \Omega}{\partial x^m} = \frac{3}{2}g^{i\alpha\dot{\beta}}u_\alpha\frac{\partial \ln \Omega}{\partial x^i}, \tag{219}$$

which cancels (216). Hence the righthand side of (215) vanishes because of (209). This establishes the required result.

APPENDIX 5 (a note to §8.2)

In Appendix 3, the contribution of the electromagnetic interaction

$$-ie\int B_i\,dx^i, \tag{220}$$

operating in a spacetime slab $t_1 \le t \le t_2$, to the wave function at point $\boldsymbol{x}_2$,

t_2, was found to be

$$-ie\int K_+(2;3)\not{B}(3)\psi_+(3)\,d\tau_3, \tag{221}$$

the spacetime integration with respect to $d\tau_3$ being taken through the slab in question. In §8.2 we are concerned with the corresponding problem when the elementary interaction $-ie\,B_i\,dx^i$ is replaced by $\frac{1}{4}\gamma_i\,dx^i$. This gives

$$\frac{1}{4}\int K_+(2;3)(\gamma_i\gamma^i)\psi_+(3)\,d\tau_3 = \int K_+(2;3)\psi_+(3)\,d\tau_3 \tag{222}$$

by the same argument as that which led to (221).

The mass interaction, unlike the electromagnetic interaction, keeps the same free-particle state ψ at all t, so that the total interaction for §8.2 is

$$\begin{aligned}\int \psi^*(2)K_+(2;3)\psi_+(3)\,d\tau_3\,d^3\mathbf{x}_2 \\ &= \int \bar{\psi}(2)\gamma_4 K_+(2;3)\psi_+(3)\,d\tau_3\,d^3\mathbf{x}_2 \\ &= \int \bar{\psi}(3)_+\psi_+(3)\,d\tau_3 = \int \bar{\psi}(3)\psi_+(3)\,d\tau_3,\end{aligned} \tag{223}$$

$\bar{\psi}_+$ being the positive-energy part of $\bar{\psi}$.

In §8.2 the wave function was written as ψ, it being implicit there that the particles under discussion had positive-energy states. Hence (223) in the notation of §8.2 would have ψ_+ replaced by ψ,

$$\int \bar{\psi}(3)\psi(3)\,d\tau_3, \tag{224}$$

which was the quoted result.

APPENDIX 6 (a note to §8.3)

With $g^i{}_{\alpha\dot{\beta};j}$ given by

$$g^i{}_{\alpha\dot{\beta};j} = \frac{\partial g^i{}_{\alpha\dot{\beta}}}{\partial x^j} + \Gamma^i{}_{jk}g^k{}_{\alpha\dot{\beta}} - L^\gamma{}_{\alpha j}g^i{}_{\gamma\dot{\beta}} - L^{\dot{\sigma}}{}_{\dot{\beta}j}g^i{}_{\alpha\dot{\sigma}}, \tag{225}$$

we are required to prove that the choice

$$L^{\gamma}{}_{\alpha j} = \frac{1}{4}\left(g^{l}{}_{\alpha\dot{\lambda}} g_{m}{}^{\gamma\dot{\lambda}} \Gamma^{m}{}_{jl} + g_{l}{}^{\gamma\dot{\lambda}} \frac{\partial g^{l}{}_{\alpha\dot{\lambda}}}{\partial x^{j}}\right), \tag{226}$$

$$L^{\dot{\sigma}}{}_{\dot{\beta} j} = \frac{1}{4}\left(g^{l}{}_{\lambda\dot{\beta}} g_{m}{}^{\lambda\dot{\sigma}} \Gamma^{m}{}_{jl} + g_{l}{}^{\lambda\dot{\sigma}} \frac{\partial g^{l}{}_{\lambda\dot{\beta}}}{\partial x^{j}}\right), \tag{227}$$

leads to

$$g^{i}{}_{\alpha\dot{\beta};j} \equiv 0. \tag{228}$$

The simplest procedure is to take advantage of the covariance of $g^{i}{}_{\alpha\dot{\beta};j}$. It follows from this that, if for a general point X we can show that $g^{i}{}_{\alpha\dot{\beta};j} \equiv 0$ for a particular choice of coordinates, then the same result holds for all coordinate systems. Choose coordinates such that for point $X(x^{j})$,

$$\frac{\partial g_{ik}}{\partial x^{l}} = 0, \qquad i, k, l = 1, 2, 3, 4, \tag{229}$$

and take the local Lorentz frame so that

$$g^{i}{}_{\alpha\dot{\beta}} = \sigma^{i}{}_{\alpha\dot{\beta}}. \tag{230}$$

In any small step $x^{j} \to x^{j} + \delta x^{j}$ we then have $\delta g_{ik} = 0$ to the first order. However, $\delta g^{i}{}_{\alpha\dot{\beta}}$ is not in general zero; $\delta g^{i}{}_{\alpha\dot{\beta}}$ is given by (92) of Chapter 8 as

$$\delta g^{i}{}_{\alpha\dot{\beta}} = \frac{1}{2} g_{k\alpha\dot{\beta}}\, \delta g^{ik} + \theta_{\alpha}{}^{\gamma} g^{i}{}_{\gamma\dot{\beta}} + \theta_{\dot{\beta}}{}^{\dot{\sigma}} g^{i}{}_{\alpha\dot{\sigma}}, \tag{231}$$

where $\delta_{\alpha}{}^{\gamma} + \theta_{\alpha}{}^{\gamma}$ can be any infinitesimal unimodular transformation, i.e., satisfying $\theta_{\gamma}{}^{\gamma} = 0$. Hence here

$$\delta g^{i}{}_{\alpha\dot{\beta}} = \theta_{\alpha}{}^{\gamma} \sigma^{i}{}_{\gamma\dot{\beta}} + \theta_{\dot{\beta}}{}^{\dot{\lambda}} \sigma^{i}{}_{\alpha\dot{\lambda}}. \tag{232}$$

Since the $\Gamma^{m}{}_{il}$ are zero, we therefore get for the variation $x^{j} \to x^{j} + \delta x^{j}$,

$$\begin{aligned} g^{i}{}_{\alpha\dot{\beta};j}\, \delta x^{j} &= \delta g^{i}{}_{\alpha\dot{\beta}} - \frac{1}{4} g_{l}{}^{\gamma\dot{\lambda}} g^{i}{}_{\gamma\dot{\beta}}\, \delta g^{l}{}_{\alpha\dot{\lambda}} - \frac{1}{4} g_{l}{}^{\lambda\dot{\sigma}} g^{i}{}_{\alpha\dot{\sigma}}\, \delta g^{l}{}_{\lambda\dot{\beta}} \\ &= \theta_{\alpha}{}^{\gamma} \sigma^{i}{}_{\gamma\dot{\beta}} + \theta_{\dot{\beta}}{}^{\dot{\lambda}} \sigma^{i}{}_{\alpha\dot{\lambda}} - \frac{1}{4} \sigma_{l}{}^{\nu\dot{\lambda}} \sigma^{i}{}_{\nu\dot{\beta}} (\theta_{\alpha}{}^{\gamma} \sigma^{l}{}_{\gamma\dot{\lambda}} + \theta_{\dot{\lambda}}{}^{\dot{\nu}} \sigma^{l}{}_{\alpha\dot{\nu}}) \\ &\quad - \frac{1}{4} \sigma_{l}{}^{\lambda\dot{\nu}} \sigma^{i}{}_{\alpha\dot{\nu}} (\theta_{\lambda}{}^{\gamma} \sigma^{l}{}_{\gamma\dot{\beta}} + \theta_{\dot{\beta}}{}^{\dot{\kappa}} \sigma^{l}{}_{\lambda\dot{\kappa}}) = 0, \end{aligned} \tag{233}$$

where we have used $\theta_\gamma{}^\gamma = 0$, $\theta_{\dot\lambda}{}^{\dot\lambda} = 0$, in the last step. This completes the proof, since (233) holds for arbitrary δx^j and therefore requires $g^i{}_{\alpha\dot\beta;j} \equiv 0$.

APPENDIX 7 (a note to §8.4)

The variations considered in (231) to (233) were with respect to a change from the point x^j to a neighboring point $x^j + \delta x^j$. Also in §8.4 we were concerned with a variation of geometry, $g_{ik} \to g_{ik} + \delta g_{ik}$. Notationally, we shall now use δ in relation to the latter variation instead of to a change of the spacetime point. A change of geometry has the effect of altering spinor derivatives, $u_{\alpha;j} \to u_{\alpha;j} + \delta u_{\alpha;j}$, according to

$$\delta u_{\alpha;j} = -\delta L^\gamma{}_{\alpha j} u_\gamma. \tag{234}$$

In §8.4 we took

$$\int g^{j\alpha\dot\beta}(u_{\dot\beta}\, \delta u_{\alpha;j} - u_\alpha\, \delta u_{\dot\beta;j} + v_\alpha\, \delta v_{\dot\beta;j} - v_{\dot\beta}\, \delta v_{\alpha;j})\sqrt{-g}\, d^4x \tag{235}$$

to be zero; here we prove it.

As a first step we calculate $\delta u_{\alpha;j}$ at a particular point. Choose coordinates so that (229), (230), hold at the point in question, and by an appropriate spin transformation arrange for

$$\frac{\partial g^{j\alpha\dot\beta}}{\partial x^k} = 0 \text{ for all } j, k, \alpha, \dot\beta \tag{236}$$

at the same point. The variation $\delta L^\gamma{}_{\alpha j}$ arising from (95) of Chapter 8,

$$\delta g^{j\alpha\dot\beta} = \frac{1}{2} g_k{}^{\alpha\dot\beta}\, \delta g^{jk}, \tag{237}$$

again at the point in question, is then given by

$$\delta L^\gamma{}_{\alpha j} = \frac{1}{4}\left(\sigma^l{}_{\alpha\dot\lambda}\sigma_m{}^{\gamma\dot\lambda}\, \delta\Gamma^m{}_{jl} + \sigma_l{}^{\gamma\dot\lambda}\frac{\delta g^l{}_{\alpha\dot\lambda}}{\partial x^j}\right). \tag{238}$$

Omitting vanishing terms

$$\sigma_m{}^{\gamma\dot\lambda}\, \delta\Gamma^m{}_{jl} = \frac{1}{2}\sigma^{m\gamma\dot\lambda}\left(\frac{\partial\delta g_{lm}}{\partial x^j} + \frac{\partial\delta g_{jm}}{\partial x^l} - \frac{\partial\delta g_{jl}}{\partial x^m}\right), \tag{239}$$

so that

$$\delta L^{\gamma}{}_{\alpha j} = \frac{1}{8}\left[\sigma^{l}{}_{\alpha\dot{\lambda}}\sigma^{m\gamma\dot{\lambda}}\left(\frac{\partial \delta g_{lm}}{\partial x^{j}} + \frac{\partial \delta g_{jm}}{\partial x^{l}} - \frac{\partial \delta g_{jl}}{\partial x^{m}}\right) + \sigma_{l}{}^{\gamma\dot{\lambda}}\sigma_{m\alpha\dot{\lambda}}\frac{\partial \delta g^{lm}}{\partial x^{j}}\right]. \tag{240}$$

The last term in the square brackets can be written in the form

$$\begin{aligned} \sigma^{l\gamma\dot{\lambda}}\sigma^{m}{}_{\alpha\dot{\lambda}}g_{il}g_{km}\frac{\partial \delta g^{ik}}{\partial x^{j}} &= \sigma^{l\gamma\dot{\lambda}}\sigma^{m}{}_{\alpha\dot{\lambda}}g_{il}\frac{\partial}{\partial x^{j}}(g_{km}\,\delta g^{ik}) \\ &= -\sigma^{l\gamma\dot{\lambda}}\sigma^{m}{}_{\alpha\dot{\lambda}}g_{il}\frac{\partial}{\partial x^{j}}(g^{ik}\,\delta g_{km}) \\ &= -\sigma^{l\gamma\dot{\lambda}}\sigma^{m}{}_{\alpha\dot{\lambda}}\frac{\partial \delta g_{lm}}{\partial x^{j}}, \end{aligned} \tag{241}$$

which cancels the $\partial(\delta g_{lm})/\partial x^{j}$ term arising from (239). We therefore get

$$\begin{aligned} \delta L^{\gamma}{}_{\alpha j} &= \frac{1}{8}\,\sigma^{l}{}_{\alpha\dot{\lambda}}\sigma^{m\gamma\dot{\lambda}}\left(\frac{\partial \delta g_{jm}}{\partial x^{l}} - \frac{\partial \delta g_{jl}}{\partial x^{m}}\right) \\ &= \frac{1}{8}\,g^{l}{}_{\alpha\dot{\lambda}}g^{m\gamma\dot{\lambda}}(\delta g_{jm;\,l} - \delta g_{jl;\,m}), \end{aligned} \tag{242}$$

where in the last line of (242) we have restored a generally covariant form for $\delta L^{\gamma}{}_{\alpha j}$. Using (234) and (242), the $\delta u_{\alpha;\,j}$ term in (235) becomes

$$\frac{1}{8}\int g^{j\alpha\dot{\beta}}g^{l}{}_{\alpha\dot{\lambda}}g^{m\gamma\dot{\lambda}}(\delta g_{jl;\,m} - \delta g_{jm;\,l})u_{\gamma}u_{\dot{\beta}}\sqrt{-g}\,d^{4}x. \tag{243}$$

Integrating by parts, setting $\delta g_{ik} = 0$ on the boundary of the region of integration, and remembering that $g^{j\alpha\dot{\beta}}{}_{;k} = 0$ for all j, k, α, $\dot{\beta}$, (243) is easily shown to be equal to

$$\frac{1}{8}\int \delta g_{jl}(u_{\gamma}u_{\dot{\beta}})_{;m}g^{j\alpha\dot{\beta}}(g^{l\gamma\dot{\lambda}}g^{m}{}_{\alpha\dot{\lambda}} - g^{m\gamma\dot{\lambda}}g^{l}{}_{\alpha\dot{\lambda}})\sqrt{-g}\,d^{4}x. \tag{244}$$

It is not hard to see that the integrand of (244) is real. The term

$$\delta g_{jl}(u_{\gamma}u_{\dot{\beta}})_{;m}g^{j\alpha\dot{\beta}}g^{l\gamma\dot{\lambda}}g^{m}{}_{\alpha\dot{\lambda}} \tag{245}$$

is easily seen to be equal to its complex conjugate: all that is needed is to take account of the symmetry of δg_{ik} and to rearrange the dummy indices after taking the complex conjugate, remembering that $g^{j\alpha\dot{\beta}}$ and $g^{j\dot{\beta}\alpha}$ are complex conjugates. For the term

$$-\delta g_{jl}(u_\gamma u_{\dot\beta})_{;m} g^{j\alpha\dot\beta} g^{m\gamma\dot\lambda} g^{l}{}_{\alpha\dot\lambda}, \tag{246}$$

we note a relation

$$g^{j\alpha\dot\beta} g^{l}{}_{\alpha\dot\lambda} = S^{jl\dot\beta}{}_{\dot\lambda} + g^{jl}\,\delta^{\dot\beta}{}_{\dot\lambda}, \tag{247}$$

in which $S^{jl\dot\beta}{}_{\dot\lambda}$ is antisymmetric in j, l. Because $S^{jl\dot\beta}{}_{\dot\lambda}$ is multiplied by the symmetric δg_{jl}, we are left with $-\delta g_{jl}(u_\gamma u_{\dot\beta})_{;m} g^{m\gamma\dot\lambda} g^{jl}\,\delta^{\dot\beta}{}_{\dot\lambda}$, i.e., with

$$-\delta g_{jl} g^{jl}(u_\gamma u_{\dot\beta})_{;m} g^{m\gamma\dot\beta}, \tag{248}$$

and this also is real.

Returning to (235), the two terms involving the spinor u are complex conjugates subtracted from each other. Since the above discussion has shown that the $\delta u_{\alpha;\,j}$ term is real, it therefore follows that the $\delta u_{\dot\beta;\,j}$ term must also be real, and that the subtraction gives zero. A similar argument shows that the two terms involving the v spinor also cancel. Hence (235) vanishes, as we required it to do.

REFERENCES

Arp, H. 1971. "NGC 7603, a galaxy connected to a companion of much larger redshift." *Astron. Lett.,* **7,** 221.

Clemence, G. M. 1947. "The relativity effect in planetary motions." *Rev. Mod. Phys.,* **19,** 361.

Courant, R., and D. Hilbert. 1962. *Methods of Mathematical Physics.* New York: Interscience.

DeWitt, B. S., and R. W. Brehme. 1960. "Radiation damping in a gravitational field." *Ann. Phys.,* **9,** 220.

Dicke, R. H. 1964. *Gravitation and Relativity.* New York: Benjamin.

Dicke, R. H., and H. M. Goldenberg. 1967. "Solar oblateness and general relativity." *Phys. Rev. Lett.,* **18,** 313.

Dirac, P. A. M. 1937. "The cosmological constants." *Nature,* **139,** 323.

Dirac, P. A. M. 1938a. "Classical theory of radiating electrons." *Proc. Roy. Soc. London,* series A, **167,** 148.

Dirac, P. A. M. 1938b. "New basis for cosmology." *Proc. Roy. Soc. London,* series A, **165,** 199.

Dirac, P. A. M. 1947. *Principles of Quantum Mechanics.* 3d ed. Oxford: Clarendon Press.

Eddington, A. S. 1930. *The Mathematical Theory of Relativity.* Cambridge, Eng.: Cambridge University Press.

Feynman, R. P. 1949. "The theory of positrons." *Phys. Rev.,* **76,** 749.

Feynman, R. P. 1950. "Mathematical formulation of the quantum theory of electromagnetic interaction." *Phys. Rev.,* **80,** 440.

Feynman, R. P., and A. R. Hibbs. 1965. *Quantum Mechanics and Path Integrals.* New York: McGraw-Hill.

Hawking, S. W., and R. Penrose. 1969. "The singularities of gravitational collapse and cosmology." *Proc. Roy. Soc. London,* series A, **314,** 529.

Hogarth, J. E. 1962. "Cosmological considerations of the absorber theory of radiation." *Proc. Roy. Soc. London,* series A, **267,** 365.

Hoyle, F., and J. V. Narlikar. 1963. "Time-symmetric electrodynamics and the arrow of time in cosmology." *Proc. Roy. Soc. London,* series A, **277,** 1.

Hoyle, F., and J. V. Narlikar. 1969. "Electrodynamics of direct interparticle action, I: The quantum-mechanical response of the universe." *Ann. Phys.,* **54,** 207.

Hoyle, F., and J. V. Narlikar. 1971. "Electrodynamics of direct interparticle action, II: Relativistic treatment of radiative processes." *Ann. Phys.,* **62,** 44.

Hoyle, F., and J. V. Narlikar. 1972. "Cosmological models in a conformally invariant gravitational theory, I and II." *Mon. Not. Roy. Astron. Soc.,* **155,** 323.

Jordan, P. 1949. "Formation of the stars and development of the galaxy." *Nature,* **164,** 637.

Muhleman, D. O., R. D. Ekers, and E. B. Fomalant. 1970. "Radio-interferometric test of the general relativistic light bending near the sun." *Phys. Rev. Lett.,* **24,** 1377.

Narlikar, J. V. 1968. "On the general correspondence between field theories and the theories of direct interparticle action." *Proc. Camb. Phil. Soc.,* **64,** 1071.

Penrose, R. 1965. "Gravitational collapse and spacetime singularities." *Phys. Rev. Lett.,* **14,** 57.

Roe, P. E. 1969. "Time-symmetric electrodynamics in Friedmann universes." *Mon. Not. Roy. Astron. Soc.,* **144,** 219.

Salam, A., and J. Strathdee. 1970. "Quantum gravity and infinities in quantum electrodynamics." *Lett. Nuovo Cimento,* **4,** 101.

Sandage, A. 1972a. "The redshift-distance relation, I." *Astrophys. J.,* **173,** 485.

Sandage, A. 1972b. "The redshift-distance relation, II." *Astrophys. J.,* **178,** 1 and 25.

Sramek, R. A. 1971. "A measurement of the gravitational deflection of microwave radiation near the sun, 1970 October." *Astrophys. J.,* **167,** L55.

Wheeler, J. A., and R. P. Feynman. 1945. "Interaction with the absorber as the mechanism of radiation." *Rev. Mod. Phys.,* **17,** 156.

Wheeler, J. A., and R. P. Feynman. 1949. "Classical electrodynamics in terms of direct interparticle action." *Rev. Mod. Phys.,* **21,** 424.

INDEX